TABLE OF CONTENTS

This handy reference is divided into five major sections: Prewar, Postwar, 1970-1989, Large Scale, and Collectibles and Paper. In the first three sections, production is listed numerically, using the item's catalogue number. In the Prewar section, equipment is also described by Gauge — either Standard, O, OO, or 2-7/8. The Gauge type is within the parentheses. To help you use this Guide, the edges of the pages are edged in black to correspond with each of the five sections.

If you cannot find an item or believe an item has been erroneously described or priced, please let us know. We are always glad to receive corrections or additions.

DEFINITIONS

This **Pocket Guide** lists prices for Prewar and Postwar Trains and Collectibles and Paper as GOOD and EXCELLENT; the prices shown for 1970-89 and Large Scale Trains are for EXCELLENT and LIKE NEW. Prices for restored pieces fall between Good and Excellent depending on the item. Mint pieces bring a substantial premium over Excellent pieces. Fair pieces bring substantially less than Good and Excellent pieces.

In the toy train field there is a great deal of concern with exterior appearance and less concern with operation. If operation is important to you, you should ask the seller if the train runs. If the seller indicates that he does not know whether the equipment operates, you should test it. Most train meets provide test tracks for this purpose.

Trains and related items are usually classified by condition. The first five categories are defined as follows:

- **FAIR** — well-scratched, chipped, dented, rusted, or warped.
- **GOOD** — scratched, dirty, with small dents.
- **VERY GOOD** — few scratches, no dents or rust, exceptionally clean.
- **EXCELLENT** — minute scratches or nicks, no dents or rust.
- **LIKE NEW** — between Excellent and Mint
- **MINT** — brand-new, absolutely unmarred, unused in original box.
- **RESTORED** — we define as professionally refinished with a color that approximates the original finish. Trim and ornamentation are present and in like new condition. The finish appears in like new condition.
- **REPRODUCTION** — is a product intended to closely resemble the original item. It may or may not be marked as such, but should be so marked. Reproductions are currently available for many desirable items.
- **NRS** (No Recorded Sales) means that we do not know the market value of the item. The item may be very scarce and bring a substantial premium over items in its general class or it may be relatively common but unnoticed. Usually NRS listings occur when an item is first reported, although we are still discovering relatively common variations that have not been previously reported. If you have information about the value of an NRS item, please write to us.
- **NA** (Not Available) means that the item is new and the market value was not available at press time.
- Numbers that have been put in parentheses by us do not appear on the items.
- **No Number** means the item may have lettering but lacks an item number.
- **(no letters)** means the item has no lettering or number on the car.

GREENBERG'S®
Pocket Price Guide and Inventory Checklist to LIONEL® TRAINS
1901-1942 & 1945-1989

By

Bruce C. Greenberg, Ph.D

With the assistance of

Charles C. Lang, Clark O'Dell, Howard LeVine, Lou Caponi, Roland LaVoie, James D. Cahill, Ed Goldman, and Robert E. Amling

Greenberg Publishing Company, Inc.
7566 Main Street
Sykesville, MD 21784
(301) 795-7447

Manufactured in the United States of America

Greenberg Publishing Company, Inc. offers the world's largest selection of Lionel, American Flyer, LGB, Ives, and other toy train publications as well as a selection of books on model and prototype railroading, dollhouse miniatures, and toys. For a copy of our current catalogue, please send a large self-addressed stamped envelope to the address above.

Greenberg Shows, Inc. sponsors the world's largest public train, dollhouse, and toy shows. They feature extravagant operating model railroads for N, HO, O, Standard, and 1 Gauges as well as a huge marketplace for buying and selling nearly all model railroad equipment. The shows also feature a large selection of dollhouses and dollhouse furnishings. Shows are currently offered in metropolitan Baltimore, Boston, Ft. Lauderdale, Cherry Hill in New Jersey, Long Island in New York, Philadelphia, Pittsburgh, and Tampa. Greenberg Auctions, a division of Greenberg Shows, Inc., offers nationally advertised auctions of toy trains and toys. Please contact Joseph Armacost at (301) 795-7447 for further information. To receive our current show listing, please send a self-addressed stamped envelope marked "Train Show Schedule" to the address above.

ISBN: 0-89778-151-1

INTRODUCTION

This **Pocket Guide** lists all major Lionel items by number only, for the years 1901-42 and 1945-early 1989. The price quoted is for the most common variety of each item. Some varieties are worth considerably more, and some rare varieties are cited. For additional information, please consult the comprehensive guides: ***Greenberg's Guide to Lionel Trains 1901-1942, Volumes I*** and ***II***, ***Greenberg's Guide to Lionel Trains 1945-1969, Volumes I*** and ***II***, and ***Greenberg's Guide to Lionel Trains: 1970-1988***, all of which are described on the rear cover.

Dates cited are catalogued dates. If there is no catalogue date, production dates are listed if known.

We have provided five columns for each item. The first two columns give the current market values for Good, Excellent, or Like New. The "Color" column is for recording the color or other distinguishing characteristics. The "Cond" column is for recording the condition of the piece. The "$" column is for recording your cost.

Values in the **Pocket Guide** are based on prices obtained at large train meets during the summer of 1989 and on prices obtained from private transactions reported by our review panel.

If you have trains to sell and you sell them to a person planning to resell them, you will NOT obtain the prices reported in this book. Rather you may expect to receive about 50 percent of these prices. For your item to be of interest to such a buyer, he must buy it for considerably less than the price listed here.

Our studies of train prices indicate that mail order prices and store prices are generally higher than those obtained at train meets. This is quite appropriate given the costs and efforts of either running a retail establishment and/or producing and distributing a price list and packing and shipping trains.

NOTES TO BUYERS

WE STRONGLY RECOMMEND THAT NOVICES DO NOT MAKE MAJOR PURCHASES WITHOUT THE ASSISTANCE OF FRIENDS WHO HAVE EXPERIENCE IN BUYING AND SELLING TRAINS.

Furthermore, extreme care should be exercised in the purchase of original prewar items which have been reproduced. Even though most reproductions were originally marked, in some cases these markings have been removed (it has been reported).

The same care needs to be exercised in the purchase of premium priced diesels. Nearly every rare diesel cab has been reproduced from an unpainted or commonly painted shell. For further information, consult ***Greenberg's Guide to Lionel Trains 1945-69, Volumes I*** and ***II***.

This **Pocket Guide** involved the collaboration of many people who generously gave their time and knowledge. Only through such cooperative ventures can studies of this kind be carried out. It is impossible for one person to put together what the factory and the marketplace dispersed to the four corners of the country (or more properly to the basements and attics of enthusiasts throughout America). I can only express my deep appreciation to my fellow enthusiasts for their dedication to toy train research. This work is not finished. Fortunately, there is always more to be learned.

Bruce C. Greenberg

SECTION I
PREWAR

	Good	Exc	Color	Cond	$
001 Steam, 4-6-4 (OO), 1938-42	120	225	___	___	___
1 Bild-A-Motor, 1928-31	60	125	___	___	___
1 Trolley, 1906-08 (Standard), 1910*	900	2000	___	___	___
002 Steam, 4-6-4 (OO), 1939-42	100	200	___	___	___
2 Bild-A-Motor, 1928-31	60	125	___	___	___
2 Countershafting, 1904-11	NRS	NRS	___	___	___
2 Trolley (Standard), 1906, 1908-16*	800	1800	___	___	___
003 Steam, 4-6-4 (OO), 1939-42	120	250	___	___	___
3 Trolley (Standard), 1906-13	1300	2700	___	___	___
004 Steam, 4-6-4 (OO), 1939-42	110	225	___	___	___
4 Electric, 0-4-0 (OO), 1928-32*	450	700	___	___	___
4 Trolley (Standard), 1906-12	3000	5000	___	___	___
4U #4 Kit form (O), 1928-29	800	1100	___	___	___
5 Electric (2-7/8") (See 100)					
5 Steam, 0-4-0 (Standard), 1906-26	400	700	___	___	___
6 Steam, 4-4-0 (Standard), 1906-23	800	1200	___	___	___
7 Steam, 4-4-0 (Standard), 1910-23*	1800	2800	___	___	___
8 Electric, 0-4-0 (Standard), 1925-32	60	125	___	___	___
8 Trolley (Standard), 1908-09*	3000	5000	___	___	___
8E Electric, 0-4-0 (Standard)	70	140	___	___	___
9 Electric, 0-4-0 (Standard), 1929*	1200	2000	___	___	___
9 Trolley (Standard), 1909	3000	5000	___	___	___
9E Electric, 0-4-0 (Standard), 1928-30*	600	1200	___	___	___
9U Electric, 0-4-0 (Standard), 1928-29	700	1350	___	___	___
10 Electric, 0-4-0 (Standard), 1925-29	75	160	___	___	___
10 Interurban (Standard), 1910-12	800	1500	___	___	___
10E Electric, 0-4-0 (Standard), 1926-36	75	165	___	___	___
011 Switches, pair (O), 1933-37	20	40	___	___	___
11 Flatcar (Standard), 1906-26	32	50	___	___	___
012 Switches, pair (O), 1927-33	12	30	___	___	___
12 Gondola (Standard), 1906-26	32	50	___	___	___
013 Pair 012 Switches, 439 Panel Board	55	125	___	___	___
13 Cattle Car (Standard), 1906-26	50	80	___	___	___
0014 Boxcar (OO), 1938-42	25	50	___	___	___
14 Boxcar (Standard), 1906-26	40	70	___	___	___
0015 Tank Car (OO), 1938-42	25	50	___	___	___
15 Oil Car (Standard), 1906-26	32	50	___	___	___
0016 Hopper Car (OO), 1938-42	35	65	___	___	___
16 Ballast Dump Car (Standard), 1906-26	85	175	___	___	___
0017 Caboose (OO), 1938-42	25	50	___	___	___
17 Caboose (Standard), 1906-26	32	50	___	___	___
18 Pullman (Standard), 1906-27	80	125	___	___	___
19 Combine (Standard), 1906-26	60	125	___	___	___
020 90° Crossover (O), 1915-42	2	5	___	___	___
020X 45° Crossover (O), 1917-42	3	7	___	___	___
20 90° Crossover (Standard)	2	5	___	___	___
20 Direct current reducer, 1906	NRS	NRS	___	___	___
20X 45° Crossover (Standard), 1928-32	3	7	___	___	___
021 Switches, pair (O), 1915-37	8	20	___	___	___

Prewar

	Good	Exc	Color	Cond	$
21 Switches, pair (Standard), 1915-25	10	20	___	___	___
21 90° Crossover (Standard), 1906	7	15	___	___	___
022 Switches, pair, Remote (O), 1938-42	15	30	___	___	___
22 Switches, pair (Standard), 1906-25	10	25	___	___	___
023 Bumper (O), 1915-33	5	10	___	___	___
23 Bumper (Standard), 1906-23	5	10	___	___	___
0024 Boxcar, P R R (OO), 1939-42	25	50	___	___	___
24 Railway Station (Standard), 1906	NRS	NRS	___	___	___
025 Bumper (O), 1928-42	5	10	___	___	___
0025 Tank Car (OO), 1939-42	25	50	___	___	___
25 Open Station, 1906	NRS	NRS	___	___	___
25 Bumper (Standard), 1927-42	7	15	___	___	___
0027 Caboose (OO), 1939-42	25	50	___	___	___
26 Passenger Bridge, 1906	NRS	NRS	___	___	___
27 Lighting Set, 1911-23	7	15	___	___	___
27 Station (Standard), 1909-12	NRS	NRS	___	___	___
28 Double Station with dome, 1909-12	NRS	NRS	___	___	___
29 Day Coach (Standard), 1909-27	375	600	___	___	___
31 Combine (Standard), 1921-25	55	80	___	___	___
32 Mail Car (Standard), 1921-25	55	80	___	___	___
32 Miniature Figures, 1909-18	—	75	___	___	___
33 Electric, 0-6-0, Early (Std.), 1912	250	600	___	___	___
33 Electric, 0-4-0 Later (Std.), 1913-24	60	125	___	___	___
34 Electric, 0-6-0 (Standard), 1912	450	900	___	___	___
34 Electric, 0-4-0 (Standard), 1913	175	400	___	___	___
35 Boulevard Lamp, 6-1/8" high, 1940-42	10	25	___	___	___
35 Pullman, first series (Std.), 1912-13	80	100	___	___	___
35 Pullman, second series (Standard)	30	50	___	___	___
35 Pullman, third series (Standard)	25	35	___	___	___
36 Observation, first series (Standard)	8O	100	___	___	___
36 Observation, second series (Standard)	35	50	___	___	___
36 Observation, third series (Standard)	30	40	___	___	___
38 Electric, 0-4-0 (Standard), 1913-24	70	125	___	___	___
41 Accessory Contactor, 1937-42	1	2	___	___	___
042 Switches, pair (O), 1938-42	15	30	___	___	___
42 Elec., 0-4-4-0, square hood (Std.), 1912	700	1300	___	___	___
42 Elec., 0-4-4-0, rnd. hood (Std.), 1913-23	175	400	___	___	___
043/43 Bild-A-Motor Gear Set, 1929	NRS	NRS	___	___	___
43 Boat, Runabout, 1933-36, 1939-41	200	350	___	___	___
0044 Boxcar (OO), 1939-42	25	50	___	___	___
0044K Boxcar Kit (OO), 1939-42	50	100	___	___	___
44 Boat, Speedster, 1935-36	250	450	___	___	___
45/045/45N Automatic Gateman, 1935-42	15	30	___	___	___
0045 Tank Car (OO), 1939-42	25	50	___	___	___
0045K Tank Car Kit (OO), 1939-42	50	100	___	___	___
0046 Hopper Car (OO), 1939-42	25	50	___	___	___
0046K Hopper Car Kit (OO), 1939-42	40	75	___	___	___
46 Crossing Gate, 1939-42	30	75	___	___	___
0047 Caboose (OO), 1939-42	25	50	___	___	___
0047K Caboose, Kit (OO), 1939-42	50	100	___	___	___
47 Crossing Gate, 1939-42	35	95	___	___	___
48W Whistle Station, 1937-42	10	20	___	___	___
49 Lionel Airport, 1937-39	70	175	___	___	___
50 Airplane, 1936	60	150	___	___	___

Prewar		Good	Exc	Color	Cond	$
50	Electric, 0-4-0 (Standard), 1924	80	150	___	___	___
50	Cardboard train, cars, acc. (O), 1943*	200	300	___	___	___
51	Steam, 0-4-0, 5 Late, 8-wheel tdr. (Std.)	750	1000	___	___	___
51	Lionel Airport, 1936, 1938	70	175	___	___	___
52	Lamp Post, 1933-41	30	60	___	___	___
53	Electric, 0-4-4-0, Early (Std.), 1912-14	800	1700	___	___	___
53	Electric, 0-4-0, Later (Standard)	450	800	___	___	___
53	Electric, 0-4-0, Latest (Std.), 1920-21	150	400	___	___	___
53	Lamp Post, 1931-42	20	35	___	___	___
54	Electric, 0-4-4-0, Early (Std.), 1912*	2100	3000	___	___	___
54	Electric, 0-4-4-0, Late (Std.), 1913-23	1300	2200	___	___	___
54	Lamp Post, 1929-35	20	35	___	___	___
55	Airplane with stand, 1937-39	150	350	___	___	___
56	Lamp Post, removable lens & cap, 24-42	15	30	___	___	___
57	Lamp Post, with street names, 1922-42	20	40	___	___	___
58	Lamp Post, 7-3/8" high, 1922-42	10	25	___	___	___
59	Lamp Post, 8-3/4" high, 1920-36	15	30	___	___	___
60/060	Telegraph Post (Standard O)	2	6	___	___	___
60	Electric, 0-4-0, F .A .O .S. (Standard)	—	2000	___	___	___
61	Electric, 0-4-4-0, F .A. O. S. (Standard)	—	2400	___	___	___
61	Lamp Post, one globe, 1914-36	15	30	___	___	___
62	Electric, 0-4-0, F. A. O. S. (Standard)	—	2000	___	___	___
62	Semaphore, 1920-32	10	25	___	___	___
63	Lamp Post, two globes, 1933-42	100	150	___	___	___
63	Semaphore, 1915-21	10	25	___	___	___
64	Lamp Post, 1940-42	10	35	___	___	___
64	Semaphore, 6-3/4" high, 1915-21	12	36	___	___	___
65	Semaphore, one-arm, 1915-26	11	30	___	___	___
65	Whistle, 1935 only	2	3	___	___	___
66	Semaphore, two-arm, 1915-26	11	30	___	___	___
66	Whistle Controller, 1936-39	2	3	___	___	___
67	Lamp Post, 1915-32	30	60	___	___	___
67	Whistle Controller, 1936-39	2	3	___	___	___
68/068	Crossing Sign	1	2	___	___	___
69/069/69N	Elec. Warning Signal, 1921-42	11	25	___	___	___
70	Outfit: (2) 62s, (1) 59, (1) 68	25	75	___	___	___
071	(6) 060 Telegraph Poles (O), 1924-42	50	100	___	___	___
71	(6) 60 Telegraph Poles (Standard)	50	100	___	___	___
0072	Switches, pair (OO), 1938-42	120	175	___	___	___
0074	Boxcar (OO), 1939-42	25	50	___	___	___
0075	Tank Car (OO), 1939-42	25	50	___	___	___
076/76	Block Signal, 1923-28	15	40	___	___	___
76	Warning Bell and Shack, 1939-42	60	150	___	___	___
0077	Caboose (OO), 1939-42	25	50	___	___	___
77/077/77N	Auto. Crossing Gate, 1923-39	15	30	___	___	___
78/078	Train Signal (Standard), 1924-32	20	50	___	___	___
79	Flashing Signal, 1928-40	30	60	___	___	___
80/080/80N	Semaphore (Std.), 1926-42	25	60	___	___	___
80	Automobile, 1912-16	600	1100	___	___	___
81	Automobile, 1912-16	600	1100	___	___	___
81	Controlling Rheostat, 1927-33	1	3	___	___	___
81	Crossing Signal, 1927-33	1	3	___	___	___
82/082/82N	Semaphore, 1927-42	30	65	___	___	___
83	Flashing Traffic Signal, 1927-42	35	75	___	___	___

Prewar		Good	Exc	Color	Cond	$
084	Semaphore, 1928-32	30	65			
84	Semaphore, 1927-32	30	65			
84	(2) Automobiles	1300	2500			
85	Telegraph Pole (Standard), 1929-42	4	10			
85	(2) Automobiles	1200	2500			
86	(6) Telegraph Poles, 1929-42	45	100			
87	Flashing Crossing Signal, 1927-42	35	75			
88	Battery Rheostat, 1915-27	1	3			
88	Rheostat Controller, 1933-42	1	3			
89	Flag Pole, 1923-34	8	20			
90	Flag Pole, 1927-42	15	35			
91	Circuit Breaker, 1930-42	7	15			
092	Signal Tower, 1923-27	40	85			
92	Floodlight Tower, 1931-42*	70	150			
93	Water Tower, 1931-42	15	30			
94	High Tension Tower, 1932-42*	70	150			
95	Controlling Rheostat, 1934-42	1	3			
96	Coal Elevator, manual, 1938-40	70	150			
097	Telegraph Set (O)	35	60			
97	Coal Elevator, 1938-42	70	150			
98	Coal Bunker, 1938-40	150	300			
99/099/99N	Train Control, 1932-42	30	75			
100	Electric Locomotive (2-7/8")*	3000	4000			
100	Trolley, 1910, 1912-15	700	1500			
100	(2) Bridge Approaches (Std.), 1920-31	10	20			
100	Wooden Gondola (2-7/8"), 1901	NRS	NRS			
101	Bridge Span, (2) Appro. (Std.), 1920-31	15	40			
101	Summer Trolley (Std.), 1910	800	1800			
102	(2) Bridge Spans, (2) Appro. (Std.), 20-31	25	60			
103	Bridge (Standard), 1913-16	10	25			
103	(3) Bridge Spans, (2) Appro. (Std.), 20-31	45	100			
104	Bridge Span (Standard), 1920-31	10	20			
104	Tunnel (Standard), 1909-14	30	75			
105	Bridge (Standard), 1911-14	10	25			
105	(2) Bridge Approaches (O), 1920-31	5	10			
106	Bridge Span, (2) Appro. (O), 1920-31	20	50			
107	Dir. current reducer, 110 V, 1923-32	NRS	NRS			
108	(2) Bridge Spans, (2) Appro. (O), 20-31	20	50			
109	(3) Bridge Spans, (2) Appro. (O), 20-32	25	60			
109	Tunnel (Standard), 1913-14	17	35			
110	Bridge Span (O), 1920-31	5	10			
111	Box of 50 Bulbs, 1920-31	—	100			
112	Gondola (Standard), 1910-26	30	50			
112	Station, 1931-35	80	175			
113	Cattle Car (Standard), 1912-26	35	55			
113	Station, 1931-34	95	200			
114	Boxcar (Standard), 1912-26	25	45			
114	Station, 1931-34	450	1000			
115	Station, 1935-42*	75	150			
116	Station, 1935-42*	450	1000			
116	Ballast Car (Standard), 1910-26	40	60			
117	Caboose (Standard), 1912-26	30	50			
117	Station, 1936-42	75	150			
118	Tunnel (O), 8" long, 1922-32	7	15			

Prewar		Good	Exc	Color	Cond	$
118L	Tunnel, 8" long, 1927	9	20			
119	Tunnel, 12" long, 1920-42	7	15			
119L	Tunnel, 12" long, 1927-33	9	20			
120	Tunnel, 17" long, 1922-27	18	35			
120L	Tunnel, 1927-42	25	45			
121	Station (Standard), 1909-16	80	200			
121	Station (Standard), 1920-26	45	100			
121X	Station (Standard), 1917-19	80	200			
122	Station (Standard), 1920-30	45	100			
123	Station (Standard), 1920-23	45	100			
123	Tunnel, 18-1/2" long (O), 1933-42	50	100			
124	Station, "Lionel City", 1920-36*	35	75			
125	Station, "Lionelville", 1923-25	50	100			
125	Track Template, 1938	1	3			
126	Station, "Lionelville", 1923-36	50	100			
127	Station, "Lionel Town", 1923-36	35	75			
128	124 Station and Terrace, 1928-35*	475	775			
128	115 Station and Terrace, 1936-42*	525	850			
129	Terrace, 1928-42*	450	700			
130	Tunnel, 26" long, 1920-36	120	250			
130L	Tunnel, 26" long, 1927-33	120	250			
131	Corner Display, 1924-28	70	200			
132	Corner Grass Plot, 1924-28	70	200			
133	Heart Shaped Plot, 1924-28	70	200			
134	Oval Shaped Plot, 1924-28	70	200			
134	Station, "Lionel City", w/stop, 1937-42	140	250			
135	Circular Plot, 1924-28	70	200			
136	Large Elevation, 1924-28	NRS	NRS			
136	Station, "Lionelville", w/stop, 1937-42	60	100			
137	Station, with stop, 1937-42	40	100			
140L	Tunnel, 37" long, 1927-32	190	400			
150	Electric, 0-4-0 Early (O), 1917	65	90			
150	Electric, 0-4-0 Late (O), 1918-25	65	90			
152	Electric, 0-4-0 (O), 1917-27	40	60			
152	Crossing Gate, 1940-42	9	20			
153	Block Signal, 1940-42	10	25			
153	Electric, 0-4-0 (O), 1924-25	60	90			
154	Electric, 0-4-0 (O), 1917-23	75	125			
154	Highway Signal, 1940-42	10	25			
155	Freight Shed, 1930-42*	90	200			
156	Electric, 4-4-4 (O), 1917-23	450	700			
156	Electric, 0-4-0 (O), 1917-23	450	700			
156	Station Platform, 1939-42	20	55			
156X	Electric, 0-4-0 (O), 1923-24	325	450			
157	Hand Truck, 1930-32	15	30			
158	Electric, 0-4-0 (O), 1923-24	60	100			
158	(2) 156s and (1) 136, 1940-42	75	150			
159	Block Actuator, 1940	10	25			
161	Baggage Truck, 1930-32*	20	50			
162	Dump Truck, 1930-32*	22	50			
163	(2) 157 Hand Trucks, one 162 Dump, and one 160 Baggage, 1930-42*	90	200			
164	Log Loader, 1940-42	70	150			
165	Magnetic Crane, 1940-42	70	150			

Prewar		Good	Exc	Color	Cond	$
166	Whistle Controller, 1940-42	2	3	___	___	___
167	Whistle Controller, 1940-42	2	3	___	___	___
167X	Whistle Controller (OO), 1940-42	2	3	___	___	___
169	Controller, 1940-42	1	3	___	___	___
170	Dir. current reducer, 220 volts, 1914-38	2	4	___	___	___
171	DC to AC Inverter, 110 volts, 1936-42	2	4	___	___	___
172	DC to AC Inverter, 229 volts, 1939-42	2	4	___	___	___
180	Pullman (Standard), 1911-21	80	90	___	___	___
181	Combine (Standard), 1911-21	80	100	___	___	___
182	Observation (Standard), 1911-21	80	100	___	___	___
184	Bungalow, illuminated, 1923-32*	25	40	___	___	___
185	Bungalow, 1923-24	17	40	___	___	___
186	(5) 185 Bungalows, 1923-32	140	300	___	___	___
186	Log Loader Outfit, 1940-41	100	200	___	___	___
187	(5) 185 Bungalows, 1923-24	140	300	___	___	___
188	Elevator and Car Set, 1938-41	90	200	___	___	___
189	Villa, illuminated, 1923-32*	55	125	___	___	___
190	Observation (Std.), 1914-18, 1923-27	80	125	___	___	___
191	Villa, illuminated, 1923-32*	55	125	___	___	___
192	Illuminated Villa Set: 191 Villa,					
189	Villa, two 184 Bungalows, 1927-32	NRS	NRS	___	___	___
193	Accessory Set, 1927-29	140	300	___	___	___
194	Accessory Set, 1927-29	140	300	___	___	___
195	Terrace, 1927-30	275	600	___	___	___
196	Accessory Set, 1927	150	300	___	___	___
200	Electric Express (2-7/8"), 1903*	4000	5000	___	___	___
200	Turntable, 1928-33*	75	150	___	___	___
200	Wooden Gondola (2-7/8"), 1902	NRS	NRS	___	___	___
200	Trailer, matches #2 Trolley (Std.)	—	3000	___	___	___
201	Steam, 0-6-0 (O), 1940-42	325	600	___	___	___
202	Summer Trolley, 1910-13	800	2400	___	___	___
203	Armored Loco, 0-4-0 (O), 1917-21	700	2000	___	___	___
203	Steam, 0-6-0 (O), 1940-42	275	400	___	___	___
204	Steam, 2-4-2 (O), 1940	30	65	___	___	___
205	Merchandise Containers, 1930-38*	120	250	___	___	___
206	Sack of Coal, 1938-42	5	15	___	___	___
208	Tool Set, 1934-42*	30	75	___	___	___
0209	Barrels, 1934-42	4	10	___	___	___
209	Wooden Barrels, 1934-42	6	15	___	___	___
210	Switches, pair (Std.), 1926, 1934-42	10	25	___	___	___
211	Flatcar (Standard), 1926-40*	45	75	___	___	___
212	Gondola (Standard), 1926-40*	45	75	___	___	___
213	Cattle Car (Standard), 1926-40*	125	175	___	___	___
214	Boxcar (Standard), 1926-40*	125	275	___	___	___
214R	Refrigerator Car (Std.), 1929-40*	300	500	___	___	___
215	Tank Car (Standard), 1926-40*	120	150	___	___	___
216	Hopper Car (Standard), 1926-38*	150	225	___	___	___
217	Caboose (Standard), 1926-40*	90	125	___	___	___
217	Lighting Set, 1914-23	NRS	NRS	___	___	___
218	Dump Car (Standard), 1926-38*	160	200	___	___	___
219	Crane (Standard), 1926-40*	75	125	___	___	___
220	Floodlight Car (Standard), 1931-40*	175	250	___	___	___
220	Switches, pair (Standard), 1926*	10	25	___	___	___
222	Switches, pair (Standard), 1926-32	20	40	___	___	___

Prewar	Good	Exc	Color	Cond	$
223 Switches, pair (Standard), 1932-42	30	75	___	___	___
224/224E Steam, 2-6-2 (O), 1938-42	60	100	___	___	___
225 222 Switches and 439 Panel, 1929-32	75	150	___	___	___
225/225E Steam, 2-6-2 (O), 1938-42	175	225	___	___	___
226/226E Steam, 2-6-4 (O), 1938-41	400	650	___	___	___
227 Steam Switcher, 0-6-0 (O), 1939-42	800	1100	___	___	___
228 Steam Switcher, 0-6-0 (O), 1939-42	800	1100	___	___	___
229 Steam Switcher, 2-4-2 (O), 1939-42	35	50	___	___	___
230 Steam Switcher, 0-6-0 (O), 1939-42	900	1500	___	___	___
231 Steam Switcher, 0-6-0 (O), 1939	900	1500	___	___	___
232 Steam Switcher, 0-6-0 (O), 1930	900	1500	___	___	___
233 Steam Switcher, 0-6-0 (O), 1940-42	900	1500	___	___	___
238/238E Steam, 4-4-2 (O), 1936-38	175	250	___	___	___
248 Electric, 0-4-0 (O), 1926-32	75	125	___	___	___
249/249E Steam, 2-4-2 (O), 1936-37	125	175	___	___	___
250E Steam, Hiawatha, 0-4-0 (O), 1935-42*	800	1300	___	___	___
250 Electric, 0-4-0 Early (O), 1926	150	200	___	___	___
250 Electric, 0-4-0 Late (O), 1934	150	200	___	___	___
251 Electric, 0-4-0 (O), 1925-32	175	250	___	___	___
251E Electric, 0-4-0 (O), 1927-32	175	250	___	___	___
252 Electric, 0-4-0 (O), 1926-32	85	125	___	___	___
252E Electric, 0-4-0 (O), 1933-35	95	135	___	___	___
253 Electric, 0-4-0 (O), 1924-32	80	125	___	___	___
253E Electric, 0-4-0 (O), 1931-36	75	125	___	___	___
254 Electric, 0-4-0 (O), 1924-32	70	150	___	___	___
254E Electric, 0-4-0 (O), 1927-34	80	175	___	___	___
255E Steam, 2-4-2 (O), 1935-36	375	500	___	___	___
256 Electric, 0-4-4-0 (O), 1924-30*	450	550	___	___	___
257 Steam, 0-4-0 (O), 1930	90	175	___	___	___
258 Steam, 2-4-0 Early (O), 1930	80	155	___	___	___
258 Steam, 2-4-2 Late (O), 1941	35	50	___	___	___
259 Steam, 2-4-2 (O), 1932	55	80	___	___	___
259E Steam, 2-4-2 (O), 1933-34, 1936-38	45	60	___	___	___
260E Steam, 2-4-2 (O), 1930-35*	300	400	___	___	___
261 Steam, 2-4-2 (O), 1931	140	200	___	___	___
261E Steam, 2-4-2 (O), 1935	160	225	___	___	___
262 Steam, 2-4-2 (O), 1931-32	100	165	___	___	___
262E Steam, 2-4-2 (O), 1933-34	110	175	___	___	___
263E Steam, 2-4-2 (O), 1936-39*	300	450	___	___	___
264E Steam, 2-4-2 (O), 1935-36	110	225	___	___	___
265E Steam, 2-4-2 (O), 1935-39	130	195	___	___	___
270 Bridge, 10" long (O), 1931-42	12	20	___	___	___
270 Lighting Set, 1915-23	NRS	NRS	___	___	___
271 (2) 270 Spans (O), 1931-33, 1935-40	25	60	___	___	___
271 Lighting Set, 1915-23	NRS	NRS	___	___	___
272 (3) 270 Spans (O), 1931-33, 1935-40	25	60	___	___	___
280 Bridge, 14" long (Standard), 1931-42	15	25	___	___	___
281 (2) Bridge Spans (Std.), 1931-33, 35-40	25	50	___	___	___
282 (3) Bridge Spans (Std.), 1931-33, 35-40	40	75	___	___	___
289E Steam, 2-4-2 (O), 1937	150	225	___	___	___
300 Elec. Trolley Car (2-7/8"), 1903-05	2000	3000	___	___	___
300 Hell Gate Bridge (Std.), 1928-42*	300	850	___	___	___
301 Batteries (2-7/8"), 1903-05	NRS	NRS	___	___	___
303 Summer Trolley (2-7/8")	1500	3000	___	___	___

		Good	Exc	Color	Cond	$
308	(5) Signs (O), 1940-42	6	15			
309	Elec. Trolley Trailer (2-7/8"), 1904-05	1700	2200			
309	Pullman (Standard), 1924-39	70	125			
310	Baggage (Standard), 1924-39	70	125			
310	Rails and Ties, complete section (2-7/8"), 1903-05	NRS	NRS			
312	Observation (Standard), 1924-39	70	125			
313	Bascule Bridge (O), 1940-42	140	375			
314	Girder Bridge (O), 1940-42	6	15			
315	Trestle Bridge (O), 1940-42	20	40			
316	Trestle Bridge (O), 1940-42	10	30			
318	Electric, 0-4-0 (Standard), 1924-32	100	225			
318E	Electric, 0-4-0 (Standard), 1926-35	110	225			
319	Pullman (Standard), 1924-27	70	125			
320	Baggage (Standard), 1925-27	70	125			
320	Turnout (2-7/8"), 1903-05	NRS	NRS			
322	Observation (Standard), 1924-27	70	125			
330	Crossing, 90° (2-7/8"), 1903-05	NRS	NRS			
332	Baggage (Standard), 1929	45	75			
337	Pullman (Standard), 1925-32	45	75			
338	Observation (Standard)	45	75			
339	Pullman (Standard), 1925-33	45	75			
340	Suspension Bridge (2-7/8"), 1903-05	NRS	NRS			
341	Observation (Standard), 1925-33	45	75			
350	Track Bumper (2-7/8"), 1903-05	20	35			
370	Jars and Plates (2-7/8"), 1903-05	NRS	NRS			
380	Electric, 0-4-0 (Standard), 1923-27	175	275			
380	Elevated Pillars (2-7/8"), 1903-05	40	75			
380E	Electric, 0-4-0 (Standard), 1926-28	175	275			
381	Electric, 4-4-4 (Standard), 1928-29*	1000	2500			
381E	Electric, 4-4-4 (Standard), 1928-36*	1000	2500			
381U	Electric, 4-4-4 (Standard), 1928-29*	1400	3500			
384	Steam, 2-4-0 (Standard), 1930-32*	275	375			
384E	Steam, 2-4-0 (Standard), 1930-32*	275	375			
385E	Steam, 2-4-2 (Standard), 1933-39*	425	635			
390	Steam, 2-4-2 (Standard), 1929*	450	750			
390E	Steam, 2-4-2 (Standard), 1929-31*	450	650			
392E	Steam, 4-4-2 (Standard), 1932-39*	650	1000			
400	Express Trail Car (2-7/8"), 1903	2000	3000			
400E	Steam, 4-4-4 (Standard), 1931-40*	1100	1800			
402	Electric, 0-4-4-0 (Standard), 1923-27	275	400			
402E	Electric, 0-4-4-0 (Std.), 1926-29	275	400			
408E	Electric, 0-4-4-0 (Std.), 1927-36*	600	1000			
412	Pullman, "California" (Std.), 1929-35*	850	1500			
413	Pullman, "Colorado" (Std.), 1929-35*	850	1500			
414	Pullman, "Illinois" (Std.), 1929-35*	975	1800			
416	Obs., "New York" (Std.), 1929-35*	850	1500			
418	Pullman (Standard), 1923-32*	135	225			
419	Combination (Standard), 1923-32*	135	225			
420	Pullman, "Faye" (Std.), 1930-40*	450	750			
421	Pullman, "Westphal" (Std.) 1930-40*	450	750			
422	Obs., "Tempel" (Standard), 1930-40*	450	750			
424	Pullman "Liberty Belle" (Std.), 1931-40*	300	475			
425	Pullman "Stephen Girard" (Std.), 31-40*	300	475			

Prewar

		Good	Exc	Color	Cond	$
426	Obs. "Coral Isle" (Std.), 1931-40*	300	475			
428	Pullman (Standard), 1926-30*	225	300			
429	Combine (Standard), 1926-30*	225	300			
430	Observation (Standard), 1926-30*	225	300			
431	Diner (Standard), 1927-32*	275	400			
435	Power Station, 1926-38*	75	150			
436	Power Station, 1926-37*	70	140			
437	Switch/Signal Tower, 1926-37*	125	250			
438	Signal Tower, 1927-39*	140	300			
439	Panel Board, 1928-42*	35	75			
440/0440/440N	Signal Bridge, 1932-42*	150	500			
440C	Panel Board, 1932-42	35	75			
441	Weighing Station (Std.), 1932-36	225	500			
442	Landscape Diner, 1938-42	60	150			
444	Roundhouse (Standard), 1932-35*	900	2500			
444-18	Roundhouse Clip, 1933	NRS	NRS			
450	Electric, Macy's, 0-4-0 (O), 1930	250	400			
450	Set: 450 & Matching (2) 605s, 606	750	1000			
455	Electric Range, 1930, 1932-33	200	500			
490	Observation (Standard), 1923-32*	135	225			
500	Dealer Display, 1927-28	NRS	NRS			
500	Elec. Derrick Car (2-7/8"), 1903-04*	3000	4000			
501	Dealer Display, 1927-28	NRS	NRS			
502	Dealer Display, 1927-28	NRS	NRS			
503	Dealer Display, 1927-28	NRS	NRS			
504	Dealer Display, 1924-28	NRS	NRS			
505	Dealer Display, 1924-28	NRS	NRS			
506	Dealer Display, 1924-28	NRS	NRS			
507	Dealer Display, 1924-28	NRS	NRS			
508	Dealer Display, 1924-28	NRS	NRS			
509	Dealer Display, 1924-28	NRS	NRS			
510	Dealer Display, 1927-28	NRS	NRS			
511	Flatcar (Standard), 1927-40	30	45			
512	Gondola (Standard), 1927-39	35	45			
513	Cattle Car (Standard), 1927-38	60	100			
514	Boxcar (Standard), 1929-40	75	100			
514	Refrigerator Car (Standard), 1927-28	150	225			
514R	Refrigerator Car (Std.), 1929-40	85	115			
515	Tank Car (Standard), 1927-40	90	125			
516	Hopper Car (Standard), 1928-40	100	150			
517	Caboose (Standard), 1927-40	35	65			
520	Floodlight Car (Standard), 1931-40	75	115			
529	Pullman (O), 1926-32	10	20			
530	Observation (O), 1926-32	10	20			
550	Miniature Figures (Std.), 1932-36*	95	150			
551	Engineer (Standard), 1932	10	20			
552	Conductor (Standard), 1932	10	20			
553	Porter (Standard), 1932	10	20			
554	Male Passenger (Standard), 1932	10	20			
555	Female Passenger (Standard), 1932	10	20			
556	Red Cap Figure (Standard), 1932	10	25			
600	Derrick Trailer (2-7/8"), 1903-04	3500	5000			
600	Pullman, Early (O), 1915-23	15	30			
600	Pullman, Late (O), 1933-42	35	75			

Prewar

		Good	Exc	Color	Cond	$
601	Observation, Late (O), 1933-42	35	75	___	___	___
601	Pullman, Early (O), 1915-23	10	25	___	___	___
602	Baggage, Lionel Lines (O), 1933-42	40	80	___	___	___
602	Baggage, N Y C (O), 1915-23	15	30	___	___	___
602	Observation, uncat. (O), 1922	10	20	___	___	___
603	Pullman, Early, uncat. (O), 1922	30	60	___	___	___
603	Pullman, Later (O), 1920-25	15	35	___	___	___
603	Pullman, Latest, 1931-36	20	35	___	___	___
604	Observation, Later (O), 1920-25	15	25	___	___	___
604	Observation, Latest (O), 1931-36	20	35	___	___	___
605	Pullman (O), 1925-32	50	100	___	___	___
606	Observation (O), 1925-32	50	100	___	___	___
607	Pullman (O), 1926-27	20	35	___	___	___
608	Observation (O), 1926-37	20	35	___	___	___
609	Pullman (O), 1937	30	60	___	___	___
610	Pullman, Early (O), 1915-25	20	50	___	___	___
610	Pullman, Late (O), 1926-30	25	50	___	___	___
611	Observation (O), 1937	30	60	___	___	___
612	Observation, Early (O), 1915-25	25	50	___	___	___
612	Observation, Late (O), 1926-30	25	50	___	___	___
613	Pullman (O), 1931-40*	50	100	___	___	___
614	Observation (O), 1931-40*	50	100	___	___	___
615	Baggage (O), 1933-40*	60	125	___	___	___
616E/616W	Diesel only (O), 1935-41	45	75	___	___	___
616E/616W	Set: 616, (2) 617s, 618	200	275	___	___	___
617	Coach (O), 1935-41	35	60	___	___	___
618	Observation (O), 1935-41	35	60	___	___	___
619	Combine (O)	85	200	___	___	___
620	Floodlight Car (O), 1937-42	20	25	___	___	___
629	Pullman (O), 1924-32	10	20	___	___	___
630	Observation, 1926-31	15	25	___	___	___
636W	Diesel only (O), 1936-41	60	100	___	___	___
636W	Set: 636W, (2) 637s, 638	200	400	___	___	___
637	Coach (O), 1936-39	50	100	___	___	___
638	Observation (O), 1936-39	50	100	___	___	___
651	Flatcar (O), 1935-40	10	20	___	___	___
652	Gondola (O), 1935-40	12	25	___	___	___
653	Hopper Car (O), 1934-40	20	30	___	___	___
654	Tank Car (O), 1934-38	12	25	___	___	___
655	Boxcar (O), 1934-42	15	30	___	___	___
656	Cattle Car (O), 1935-40	25	40	___	___	___
657	Caboose (O), 1934-42	10	20	___	___	___
659	Dump Car (O), 1935-42	25	40	___	___	___
700	Electric, 0-4-0 (O), 1913-16	250	400	___	___	___
700	Window Display (2-7/8"), 1903-05	NRS	NRS	___	___	___
700E	Steam, 4-6-4 Scale Hudson, "5344" (O), 1937-42*	1900	3500	___	___	___
700K	Steam, 4-6-4 unbuilt kit (O), 1938-42	3500	5000	___	___	___
701	Electric, 0-4-0 (O), 1913-16	350	500	___	___	___
701	Steam, 0-6-0 (See 708)					
702	Baggage (O), 1917-21	90	200	___	___	___
703	Electric, 4-4-4 (O), 1913-16	1300	2000	___	___	___
706	Electric, 0-4-0 (O), 1913-16	300	500	___	___	___

Item	Description	Good	Exc	Color	Cond	$
708	Steam, 0-6-0, "8976" on boiler front (O), 1939-42*	1900	3000			
710	Pullman (O), 1924-34	75	175			
711	R.C. Switches, pair (O72), 1935-42	75	150			
712	Observation (O), 1924-34	75	175			
714	Boxcar, scale (O), 1940-42*	300	500			
714K	Boxcar, unbuilt kit (O), 1940-42	—	800			
715	Shell Tank Car, scale (O), 1940-42*	275	450			
715K	Tank Car, unbuilt kit (O), 1940-42	—	650			
716	Hopper Car, scale (O), 1940-42*	400	600			
716K	Hopper, unbuilt kit (O), 1940-42	—	850			
717	Caboose (O), 1940-42*	325	450			
717K	Caboose, unbuilt kit (O), 1940-42	—	650			
720	90° Crossing (O72), 1935-42	7	15			
721	Manual Switches, pair (O72), 1935-42	45	100			
730	90° Crossing (O72), 1935-42	20	35			
731	R.C. Switches, pair, T-rail (O72), 35-42	85	175			
752E/752W	Set: 752, (2) 753s, 754 (O), 1934-41*	600	800			
752E	Set: 752, 753, 754 (O), 1934-41*	400	650			
752E	Diesel only	130	200			
753	Coach (O), 1936-41	75	125			
754	Observation (O), 1936-41	75	125			
760	16-piece Curved Track (O72), 1935-42	25	50			
761	Curved Track (O72), 1934-42	1	2			
762	Straight Track (O72), 1934-42	1	2			
762	Insulated Straight Track (O72), 1934-42	2	3			
763E	Steam, 4-6-4 (O), 1937-42	1000	2750			
771	Curved Track, T-rail (O72), 1935-42	3	6			
772	Straight Track, T-rail (O72), 1935-42	4	10			
773	Fishplate Outfit (O72), 1936-42	25	30			
782	Milwaukee Rd Combine (O), 1935-41*	185	400			
783	Milwaukee Rd Coach (O), 1935-41*	185	400			
784	Milw. Rd Observation (O), 1935-41*	185	400			
792	Rail Chief Combine (O), 1937-41*	300	800			
793	Rail Chief Coach (O), 1937-41*	300	800			
794	Rail Chief Observation (O), 1927-41*	300	800			
800	Boxcar (O), 1915-26	10	35			
800	Boxcar (2-7/8"), 1904-05*	2000	3000			
801	Caboose (O), 1915-26	15	35			
802	Stock Car (O), 1915-26	20	40			
803	Hopper Car, Early (O), 1923-28	20	30			
803	Hopper Car, Late (O), 1929-34	20	25			
804	Tank Car (O), 1923-28	20	25			
805	Boxcar (O), 1927-34	20	25			
806	Stock Car (O), 1927-34	20	30			
807	Caboose (O), 1927-40	20	25			
809	Dump Car (O), 1930-41	20	35			
810	Crane (O), 1930-42	50	100			
811	Flatcar (O), 1926-40	30	55			
812	Gondola (O), 1926-42	20	40			
813	Stock Car (O), 1926-42	30	65			
814	Boxcar (O), 1926-42	20	75			
814R	Refrigerator Car (O), 1929-42	50	125			

Prewar

		Good	Exc	Color	Cond	$
815	Tank Car (O), 1926-42	25	55	___	___	___
816	Hopper Car (O), 1927-42	35	80	___	___	___
817	Caboose (O), 1926-42	25	55	___	___	___
820	Boxcar (O), 1915-26	25	40	___	___	___
820	Floodlight Car (O), 1931-42	35	85	___	___	___
821	Stock Car (O), 1915-16, 1925-26	45	75	___	___	___
822	Caboose (O), 1915-26	35	50	___	___	___
831	Flatcar (O), 1927-34	15	25	___	___	___
840	Industrial Power Station, 1928-40*	500	2200	___	___	___
900	Ammunition Car (O), 1917-21	60	125	___	___	___
900	Box Trail Car (2-7/8"), 1904-05	2000	3000	___	___	___
901	Gondola (O), 1919-27	20	35	___	___	___
902	Gondola (O), 1927-34	15	25	___	___	___
910	Grove of Trees, 1932-42	70	150	___	___	___
911	Country Estate	175	350	___	___	___
912	Suburban Home	175	350	___	___	___
913	Landscaped Bungalow, 1940-42	140	275	___	___	___
914	Park Landscape, 1932-35	90	200	___	___	___
915	Tunnel, 1932, 1934-35	100	250	___	___	___
916	Tunnel, 29-1/4" long, 1935	70	150	___	___	___
917	Scenic Hillside, 1932-36	90	200	___	___	___
918	Scenic Hillside, 1932-36	90	200	___	___	___
919	Park Grass, bag, 1932-42	7	15	___	___	___
920	Village, 1932-33	600	1500	___	___	___
921	Scenic Park, three pieces, 1932-33	1000	2500	___	___	___
921C	Park Center, 1932-33	400	1000	___	___	___
922	Terrace, 1932-36	80	150	___	___	___
923	Tunnel, 40-1/4" long, 1933-42	70	200	___	___	___
924	Tunnel, 30" long (O72), 1935-42	50	125	___	___	___
925	Lubricant, 1935-42	1	2	___	___	___
927	Flag Plot, 1937-42	60	125	___	___	___
1000	Passenger Car (2-7/8"), 1905*	3500	5000	___	___	___
1000	Trailer (2-7/8")	NRS	NRS	___	___	___
1010	Electric, 0-4-0 (O), 1931-32	45	75	___	___	___
1010	Trailer, matches 10 Interurban	600	1400	___	___	___
1011	Pullman (O), 1931-32	10	30	___	___	___
1012	Station, 1932	20	35	___	___	___
1015	Steam, 0-4-0 (O), 1931-32	60	90	___	___	___
1017	Winner Station, 1933	15	40	___	___	___
1019	Observation (O), 1931-32	25	50	___	___	___
1020	Baggage (O), 1931-32	50	100	___	___	___
1021	90° Crossover (O27), 1932-42	1	2	___	___	___
1022	Tunnel, 18-3/4" long (O), 1935-42	7	15	___	___	___
1023	Tunnel, 19" long, 1934-42	10	15	___	___	___
1024	Switches, pair (O27), 1937-42	2	5	___	___	___
1025	Bumper (O27), 1940-42	6	15	___	___	___
1027	Transformer in Tin Station, 1934	15	40	___	___	___
1028	Transformer, 40 watts, 1939	2	4	___	___	___
1030	Electric, 0-4-0 (O), 1932	60	90	___	___	___
1035	Steam, 0-4-0 (O), 1932	65	95	___	___	___
1045	Watchman, 1938-42	10	25	___	___	___
1050	Passenger Car Trailer (2-7/8"), 1905	3500	5000	___	___	___
1100	Handcar, Mickey Mouse (O) 1935-37*	400	500	___	___	___
1100	Trailer, matches 101 (Standard)	NRS	NRS	___	___	___

Prewar

		Good	Exc	Color	Cond	$
1103	Handcar, Peter Rabbit (O), 1935-37*	400	850	____	____	____
1105	Handcar, Santa Claus (O), 1935-36*	575	900	____	____	____
1107	Transformer in Tin Station, 1933	15	40	____	____	____
1107	Handcar, Donald Duck (O), 1936-37*	425	700	____	____	____
1121	Switches, pair (O27), 1937-42	15	25	____	____	____
1506L	Steam, 0-4-0 (O), 1933-34	90	115	____	____	____
1506M	Steam, 0-4-0 (O), 1935	225	400	____	____	____
1508	Steam, 0-4-0, C.V. with Mickey	275	400	____	____	____
1511	Steam, 0-4-0 (O), 1936-37	100	150	____	____	____
1512	Gondola (O), 1931-33, 1936-37	6	12	____	____	____
1514	Boxcar (O), 1931-37	6	12	____	____	____
1515	Tank Car (O), 1933-37	6	12	____	____	____
1517	Caboose (O), 1931-37	6	12	____	____	____
1518	Mickey Mouse Diner (O), 1935	35	75	____	____	____
1519	Mickey Mouse Band (O), 1935	35	75	____	____	____
1520	Mickey Mouse Animal (O), 1935	35	75	____	____	____
1536	Circus: 1508, 1509, 1518, 1519, 1520	550	800	____	____	____
1550	Switches, pair, Windup, 1933-37	2	5	____	____	____
1555	90° Crossover, Windup, 1933-37	1	2	____	____	____
1560	Station, 1933-37	12	20	____	____	____
1569	Accessory Set, eight pieces, 1933-37	20	45	____	____	____
1588	Steam, 0-4-0 (O), 1936-37	100	150	____	____	____
•1630	Pullman (O), 1938-42	15	25	____	____	____
•1631	Observation (O), 1938-42	15	25	____	____	____
1651E	Electric, 0-4-0 (O), 1933	100	150	____	____	____
1661E	Steam, 2-4-0 (O), 1933	65	100	____	____	____
1662	Steam, 0-4-0 (O), 1940-42	130	200	____	____	____
1663	Steam, 0-4-0 (O), 1940-42	160	250	____	____	____
1664/1664E	Steam, 2-4-2 (O), 1938-42	45	65	____	____	____
1666/1666E	Steam, 2-6-2 (O), 1938-42	45	75	____	____	____
1668/1668E	Steam, 2-6-2 (O), 1937-41	60	100	____	____	____
1673	Coach (O), 1936-37	20	35	____	____	____
1674	Pullman (O), 1936-37	20	35	____	____	____
1675	Observation (O), 1936-37	20	35	____	____	____
1677	Gondola (O), 1933-35	10	30	____	____	____
1679	Boxcar (O), 1933-42	5	15	____	____	____
1680	Tank Car (O), 1933-42	5	15	____	____	____
1681E	Steam, 2-4-0 (O), 1934-35	30	70	____	____	____
1681	Steam, 2-4-0 (O), 1934-35	30	70	____	____	____
1682	Caboose (O), 1933-42	5	15	____	____	____
1684	Steam, 2-4-2 (O), 1942	30	50	____	____	____
1685	Coach, uncatalogued (O)	100	200	____	____	____
1686	Baggage, uncatalogued (O)	100	200	____	____	____
1687	Observation, uncatalogued (O)	100	200	____	____	____
1688/1688E	Steam, 2-4-2 (O), 1936	30	50	____	____	____
1689E	Steam, 2-4-2 (O), 1936-37	50	70	____	____	____
1690	Pullman (O), 1933-34	8	15	____	____	____
1691	Observation, uncatalogued (O)	8	15	____	____	____
1692	Pullman, uncatalogued (O), 1939	15	30	____	____	____
1693	Observation, uncatalogued (O)	15	30	____	____	____
1700E	Diesel, power unit only (O)	20	40	____	____	____
1700E	Set: 1700, (2) 1701s, 1702 (O), 35-37	55	125	____	____	____
1701	Coach (O), 1935-37	10	25	____	____	____
1702	Observation (O), 1935	10	25	____	____	____

Prewar

		Good	Exc	Color	Cond	$
1703	Observation w/hooked coupler, uncat.	35	75	___	___	___
1717	Gondola, uncatalogued (O), 1933-40	10	20	___	___	___
1717X	Gondola, uncatalogued (O), 1940	10	30	___	___	___
1719	Boxcar, uncatalogued (O), 1933-40	15	25	___	___	___
1719X	Boxcar, uncatalogued (O), 1941-42	15	25	___	___	___
1722	Caboose, uncatalogued (O), 1933-42	15	25	___	___	___
1722X	Caboose, uncatalogued (O), 1939-40	8	20	___	___	___
1766	Pullman (Standard), 1934-40*	300	500	___	___	___
1767	Baggage (Standard), 1934-40*	300	500	___	___	___
1768	Observation (Standard), 1934-40*	300	500	___	___	___
1811	Pullman (O), 1933-37	10	30	___	___	___
1812	Observation (O), 1933-37	25	50	___	___	___
1813	Baggage (O)	50	100	___	___	___
1816/1816W	Diesel (O), 1935-37	65	200	___	___	___
1817	Coach (O), 1935-37	20	40	___	___	___
1818	Observation (O), 1935-37	20	40	___	___	___
1835E	Steam, 2-4-2 (Standard), 1934-39	400	600	___	___	___
1910	Electric, 0-6-0 Early (Std.), 1910-11	750	1600	___	___	___
1910	Electric, 0-6-0 Late (Std.), 1912	450	1000	___	___	___
1910	Pullman (Standard), 1909-10	800	1500	___	___	___
1911	Electric, 0-4-0 Early (Std.), 1910-12	1000	2000	___	___	___
1911	Electric, 0-4-0 Late (Std.), 1913	700	1200	___	___	___
1911	Elec., 0-4-4-0 Special (Std.), 1911-12	1000	2000	___	___	___
1912	Electric, 0-4-4-0 (Std.), 1910-12	1500	2700	___	___	___
1912	Elec., 0-4-4-0 Special (Std.), 1911*	2000	5000	___	___	___
2200	Trailer, matches 202 (Standard)	1100	2500	___	___	___
2600	Pullman (O), 1938-42	45	100	___	___	___
2601	Observation (O), 1938-42	45	100	___	___	___
2602	Baggage (O), 1938-42	65	125	___	___	___
2613	Pullman (O), 1938-42*	90	200	___	___	___
2614	Observation (O), 1938-42*	90	200	___	___	___
2615	Baggage (O), 1938-42*	115	225	___	___	___
2620	Floodlight (O), 1938-42	18	45	___	___	___
2623	Pullman (O), 1941-42	90	200	___	___	___
2624	Pullman (O), 1941-42	700	1500	___	___	___
2630	Pullman (O), 1938-42	15	30	___	___	___
2631	Observation (O), 1938-42	15	30	___	___	___
2640	Pullman, illuminated (O), 1938-42	20	35	___	___	___
2641	Observation, illuminated (O), 1938-42	20	35	___	___	___
2642	Pullman (O), 1941-42	15	30	___	___	___
2643	Observation (O), 1941-42	15	30	___	___	___
2651	Flatcar (O), 1938-42	15	35	___	___	___
2652	Gondola (O), 1938-41	10	30	___	___	___
2653	Hopper Car (O), 1938-42	20	45	___	___	___
2654	Tank Car (O), 1938-42	10	35	___	___	___
2655	Boxcar (O), 1938-39	20	55	___	___	___
2656	Stock Car (O), 1938-39	30	60	___	___	___
2657	Caboose (O), 1940-41	10	20	___	___	___
2657X	Caboose (O), 1940-41	10	20	___	___	___
2659	Dump Car (O)	25	40	___	___	___
2660	Crane (O), 1938-42	21	45	___	___	___
2672	Caboose (O), 1942	10	20	___	___	___
2677	Gondola (O)	9	18	___	___	___
2679	Boxcar (O)	9	20	___	___	___

Prewar	Good	Exc	Color	Cond	$
2680 Tank Car (O), 1938-42	9	20	___	___	___
2682 Caboose (O), 1938-42	8	20	___	___	___
2682X Caboose (O), 1938-42	8	20	___	___	___
2717 Gondola (O), uncatalogued	14	30	___	___	___
2719 Boxcar (O), uncatalogued	15	30	___	___	___
2722 Caboose (O), uncatalogued	14	30	___	___	___
2755 Tank Car (O), 1941-42	25	60	___	___	___
2757 Caboose (O), 1941-42	20	30	___	___	___
2757X Caboose (O), 1941-42	20	30	___	___	___
2758 Automobile Boxcar (O), 1941-42	25	35	___	___	___
2810 Crane (O), 1938-42	90	175	___	___	___
2811 Flatcar (O), 1938-42	60	125	___	___	___
2812 Gondola (O), 1938-42	30	65	___	___	___
2813 Stock Car (O), 1938-42	100	175	___	___	___
2814 Boxcar (O), 1938-42	75	125	___	___	___
2814R Refrigerator Car (O), 1938-42	160	300	___	___	___
2815 Tank Car (O), 1938-42	45	95	___	___	___
2816 Hopper Car (O), 1935-42	45	110	___	___	___
2817 Caboose (O), 1936-42	35	75	___	___	___
2820 Floodlight Car (O), 1938-42	75	150	___	___	___
2954 Boxcar (O), 1940-42*	200	450	___	___	___
2955 Sunoco Tank Car (O), 1940-42*	200	500	___	___	___
2956 Hopper Car (O), 1940-42*	200	425	___	___	___
2957 Caboose (O), 1940-42*	200	425	___	___	___
3300 Trailer	1400	3000	___	___	___
3651 Operating Lumber Car (O), 1939-42	10	20	___	___	___
3652 Operating Gondola (O), 1939-42	17	40	___	___	___
3659 Operating Dump Car (O), 1939-42	11	20	___	___	___
3811 Oper. Lumber Dump Car (O), 1939-42	25	50	___	___	___
3814 Oper. Merchandise Car (O), 1929-42	90	200	___	___	___
3859 Operating Dump Car (O), 1938-42	20	50	___	___	___
5344 (See 700E)					
8976 (See 227, 228, 229, 230, 706, 708)					
A Miniature Motor, 1904	50	100	___	___	___
A Transformer, 40 or 60 watts, 1927-37	1	2	___	___	___
B New Departure Motor, 1906-16	50	100	___	___	___
B Transformer, 50 or 75 watts, 1916-38	1	2	___	___	___
C New Departure Motor, 1906-16	50	100	___	___	___
D New Departure Motor, 1906-14	50	100	___	___	___
E New Departure Motor, 1906-14	50	100	___	___	___
F New Departure Motor, 1906-14	50	100	___	___	___
G Battery Fan Motor, 1906-14	50	100	___	___	___
K Power Motor, 1904-06	50	100	___	___	___
K Transformer, 150 or 200 watts	2	5	___	___	___
L Power Motor, 1905	50	100	___	___	___
L Transformer, 50 or 75 watts	1	2	___	___	___
M Battery Motor, 1915-20	30	75	___	___	___
N Transformer, 50 watts	1	2	___	___	___
Q Transformer, 50 watts	1	2	___	___	___
Q Transformer, 75 watts	7	12	___	___	___
R Battery Motor, 1915-20	30	75	___	___	___
R Transformer, 100 watts, 1938-42	10	20	___	___	___
S Transformer, 50 watts	1	2	___	___	___
S Transformer, 80 watts	7	12	___	___	___

Prewar

		Good	Exc	Color	Cond	$
T	Transformer, 75, 100, or 150 watts	1	3	___	___	___
U	Transformer, Alladin	1	2	___	___	___
V	Transformer, 150 watts, 1939-42	25	45	___	___	___
W	Transformer, 75 watts	1	3	___	___	___
Y	Battery Motor, 1915-20	40	80	___	___	___
Z	Transformer, 250 watts, 1939-42	35	60	___	___	___

Other transformers and rheostats made by Lionel:

		Good	Exc	Color	Cond	$
106	Rheostat, 1911-14	3	9	___	___	___
1029	25 watts, 1936	1	2	___	___	___
1030	40 watts, 1935-38	2	3	___	___	___
1031	Rheosat, circa 1938	2	3	___	___	___
1036	Rheostat, circa 1941	2	3	___	___	___
1037	40 watts, 1940-42	1	2	___	___	___
1038	Rheostat, circa 1940	2	3	___	___	___
1039	35 watts, 1937-40	1	2	___	___	___
1040	60 watts, 1937-39	5	10	___	___	___
1041	60 watts, 1939-42	5	10	___	___	___

TRACK, LOCKONS, AND CONTACTORS

	Good	Exc	Color	Cond	$
O Straight	.20	.50	___	___	___
O Curve	.15	.40	___	___	___
O72 Straight	1	2	___	___	___
O72 Curve	1	2	___	___	___
O27 Straight	.10	.30	___	___	___
O27 Curve	.10	.20	___	___	___
Standard Straight	.50	1	___	___	___
Standard Curve	.50	1	___	___	___
O Gauge Lockon	.10	.25	___	___	___
Standard Gauge Lockon	.25	.50	___	___	___
UTC Lockon	.25	.75	___	___	___
145C Contactor	.50	1	___	___	___
153C Contactor	.50	1.50	___	___	___

SECTION II
POSTWAR

	Good	Exc	Color	Cond	$
011-11 Fiber Pins (O), 1946-50	.03	.05			
011-11 Insulating Pins (O), 1940-60	.03	.05			
011-43 Insulating Pins dozen (O), 1961	.40	1			
020 90° Crossover (O), 1945-61	1.50	4			
020X 45° Crossover (O), 1946-59	1.50	4			
022 R.C. Switches, pair (O), 1945-49	25	55			
022-500 Adapter Set (O), 1957-61	1	2			
022A R.C. Switches, pair (O)	—	100			
025 Bumper (O), 1946-47	3	8			
026 Bumper, 1948-50	3	8			
027C-1 Track Clips, dz. (O27), 1947, 1949	.50	1			
30 Water Tower, 1947-50	25	80			
31 Curved Track (Super O), 1957-66	.30	.60			
31-7 Power Blade Con. (Super O), 1957-61	—	.25			
31-15 Ground Rail Pin (Super O), 1957-66	—	.75			
31-45 Power Blade Con. (Super O), 1961-66	—	.75			
32 Straight Track (Super O), 1957	.35	.75			
32-10 Insulating Pin (Super O), 1957-60	—	.50			
32-20 Power Blade Ins. (Super O), 1957-60	—	.10			
32-25 Insulating Pin (Super O)	—	.10			
32-30 Ground Pin (Super O)	—	.10			
32-31 Power Pin (Super O)	—	.10			
32-32 Insulating Pin (Super O)	—	.10			
32-33 Ground Pin (Super O)	—	.10			
32-34 Power Pin (Super O)	—	.10			
32-45 Pow. Blade Ins., dz. (Sup. O), 1961-66	—	.75			
32-55 Insulating Pin (Super O), 1961-66	—	.75			
33 Half Curved Track (Super O), 1957-66	.25	.85			
34 Half Straight Track (Super O), 1957-66	.25	.75			
35 Boulevard Lamp, 1945-49	8	20			
36 Remote Control Set (Super O), 1957-60	1	4			
37 Uncoupling Track Set (Super O), 1957-66	1	4			
38 Water Tower, 1946-47	50	250			
38 Acc. Adapter Track (Super O), 1957-61	1	4			
39 Operating Set (Super O), 1957	1	4			
39-25 Operating Set (Super O), 1961-66	1	4			
40 Hookup Wire, 1950-51, 1953-63	1	3			
40-25 Conductor Wire, 1956-59	1	4			
40-50 Cable Reel, 1960-61	1	3			
41 Contactor	.40	.80			
41 U. S. Army Switcher, 1955-57	45	95			
42 Picatinny Arsenal Switcher, 1957	100	200			
042/42 Manual Switches, pair (O), 1946-59	10	35			
43 Power Track, 1959-66	1.50	3			
44-80 Missiles, 1959-60	3	7			
44 U. S. Army Mobile Launcher, 1959-62	60	150			
45 U. S. Mobile Launcher Switcher, 1960-62	75	250			
45 Automatic Gateman, 1946-49	15	30			
45N Automatic Gateman, 1945	15	30			
48 Ins. Straight Track (Super O), 1957-66	.75	1.50			

Postwar

		Good	Exc	Color	Cond	$
50	Lionel Gang Car, 1954-64	30	55	___	___	___
51	Navy Yard Switcher, 1956-57	65	150	___	___	___
52	Fire Car, 1958-61	100	225	___	___	___
53	Rio Grande Snowplow, 1957-60					
	(A) Backwards "a" in Rio Grande	100	300	___	___	___
	(B) Correctly printed "a"	200	400	___	___	___
54	Ballast Tamper, 1958-61, 1966, 1968-69	100	200	___	___	___
54-6446	N & W Quad Hopper, 1954	10	30	___	___	___
54-6446	N & W Cement, gray	10	30	___	___	___
55-150	Ties, 1957-60	1.50	3	___	___	___
55	Tie-jector, 1957-61	100	175	___	___	___
56	Lamp Post, 1946-49	10	35	___	___	___
56	M & St L Mine Transport, 1958	175	500	___	___	___
57	A E C Switcher, 1959-60	250	675	___	___	___
58	Lamp Post, 1946-60	8	25	___	___	___
58	G N Snowplow 2-4-2, 1959-61	275	600	___	___	___
59	Minuteman Switcher, 1962-63	225	550	___	___	___
60	Lionelville Trolley, 1955-58	100	275	___	___	___
61	Ground Lockon (Super O), 1957-60	.25	.50	___	___	___
62	Power Lockon (Super O), 1957-66	.25	.50	___	___	___
64	Street Lamp, 1945-49	10	35	___	___	___
65	Lionel Lines Handcar, 1962-66	100	325	___	___	___
68	Executive Inspection Car, 1958-61	100	250	___	___	___
69	Lionel Maintenance Car, 1960-62	150	300	___	___	___
70	Yard Light, 1949-50	10	35	___	___	___
71	Lamp Post, 1949-59	2	10	___	___	___
75	Tear Drop Lamp, 1961-63	5	15	___	___	___
76	Boulevard Street Lamp, 1955-56, 1968-69	2	4	___	___	___
88	Controller, 1946-60	.50	1	___	___	___
89	Flagpole, 1956-58	10	30	___	___	___
90	Controller	.25	.75	___	___	___
91	Circuit Breaker, 1957-60	1	3	___	___	___
92	Circuit Breaker Controller, 1959-66, 68-69	.50	1	___	___	___
93	Water Tower, 1946-49	7	25	___	___	___
97	Coal Elevator, 1946-50	50	135	___	___	___
100	Multivolt-DC/AC, trans., 1958-66	NRS	NRS	___	___	___
109	Trestle Set, 1961	NRS	NRS	___	___	___
110	Trestle Set, 1955-69	3	15	___	___	___
111	Trestle Set, 1956-69	3	8	___	___	___
111-100	Trestle Piers, 1960-63	1	3	___	___	___
112	R.C. Switches, pair (Super O), 1957-66	20	75	___	___	___
114	Newsstand with horn, 1957-59	20	75	___	___	___
115	Passenger Station, 1946-49	100	200	___	___	___
118	Newsstand with whistle, 1958	20	75	___	___	___
119	Landscaped Tunnel, 1957	2	5	___	___	___
120	90° Crossing (Super O), 1957-66	1	3	___	___	___
121	Landscaped Tunnel, 1959-66	2	5	___	___	___
122	Landscaped Tunnel, 1968	2	5	___	___	___
123	Lamp Assortment, 1955-59	NRS	NRS	___	___	___
123-60	Lamp Assortment, 1960-63	NRS	NRS	___	___	___
125	Whistle Shack, 1950-55	7	30	___	___	___
128	Animated Newsstand, 1957-60	50	100	___	___	___
130	60° Crossing (Super O), 1957-61	1	4	___	___	___
131	Curved Tunnel, 1959-66	2	5	___	___	___

		Good	Exc	Color	Cond	$
132	Passenger Station, 1949-55	11	30			
133	Passenger Station, 1957-66	11	25			
137	Passenger Station, 1946 (See Prewar section)					
138	Water Tower, 1953-57	35	70			
140	Automatic Banjo Signal, 1954-66	10	20			
142	Manual Switches, pr. (Sup. O), 1957-66	10	20			
145C	Contactor, 1950-60	.50	1.50			
145	Automatic Gateman, 1950-66	15	30			
147	Whistle Controller, 1961-66	.50	1			
148	Dwarf Trackside Signal, 1957-60	15	50			
150	Telegraph Pole Set, 1947-50	5	25			
151	Automatic Semaphore, 1947-69	10	25			
152	Automatic Crossing Gate, 1945-48	10	20			
153	Auto. Block Control, Signal, 1945-69	10	20			
153C	Contactor	.50	2			
154	Auto Highway Signal, 1945-69	7	20			
155	Blinking Light Signal w/bell, 1955-57	10	25			
156	Station Platform, 1946-51	15	40			
157	Station Platform, 1952-59	6	20			
160	Unloading Bin, 1952	.25	1			
161	Mail Pickup Set, 1961-63	20	75			
163	Single Target Block Signal, 1961-69	7	20			
164	Log Loader, 1946-50	65	150			
165	Crane, 1946-47	70	150			
167	Whistle Controller, 1945-46	1	3			
175	Rocket Launcher, 1958-60	50	150			
175-50	Extra Rocket, 1959-60	1.50	3			
182	Magnetic Crane, 1946-49	70	150			
192	Operating Control Tower, 1959-60	60	150			
193	Industrial Water Tower, 1953-55	30	75			
195	Floodlight Tower, 1957-69	20	30			
195-75	Eight-Bulb Extension, 1958-60	6	10			
196	(A) Smoke Pellets, 1946	—	20			
196	(B) Smoke Pellets, 1947	—	20			
197	Rotating Radar Antenna, 1958-59	12	50			
199	Microwave/Relay Tower, 1958-59	18	65			
202	U P Alco A unit, 1957	25	75			
204	Santa Fe Alco AA units, 1957	50	125			
205	M P Alco AA units, 1957-58	50	125			
206	Artificial Coal, large bag, 1946-68	—	6			
207	Artificial Coal, small bag	—	4			
208	Santa Fe Alco AA units, 1958-59	50	150			
209	New Haven Alco AA units, 1958	100	350			
209	Wooden Barrels, set of four, 1946-50	—	3			
210	Texas Special Alco AA units, 1958	50	150			
211	Texas Special Alco AA units, 1962-66	60	125			
212	U. S. Marine Corps Alco A, 1958-59	50	135			
212	Santa Fe Alco AA units, 1964-66	50	150			
212T	U. S. M. C. Dummy, 1958-59	175	375			
213	Railroad Lift Bridge, 1950			Not Manufactured		
213	M & St L Alco AA units, 1964	50	150			
214	Plate Girder Bridge, 1953-69	2	5			

		Good	Exc	Color	Cond	$
215	Santa Fe Alco AB units, 1965					
	(A) AB units	50	150			
	(B) Double A units	50	100			
216	(A) Burlington Alco A unit, 1958	75	250			
216	(B) M & St L Alco A unit	75	150			
217	B & M Alco AB units, 1959	50	150			
218	Santa Fe Alco units, 1959-63					
	(A) Double A units	60	125			
	(B) AB units	60	125			
219	M P Alco AA units, circa 1959	50	125			
220	Santa Fe Alco A unit, 1961	50	125			
221	2-6-4, 221T/221W Tender, 1946-47	50	115			
221	Rio Grande Alco E unit, 1963-64	35	75			
221	U. S. Marine Corps Alco A unit	75	225			
221	Santa Fe Alco E-unit	75	225			
222	Rio Grande Alco A unit, 1962	35	75			
223	218C Santa Fe Alco AB units, 1963	50	150			
224	Steam 2-6-2, 2466T/2466W Tdr., 45-46	60	100			
224	U. S. Navy Alco AB units, 1960	75	175			
225	C & O Alco A unit, 1960	40	100			
226	B & M Alco AB units, 1960	50	125			
227	C N Alco A unit, 1960	50	125			
228	C N Alco A unit, 1960	50	100			
229	M & St L Alco AB units, 1961-62					
	(A) 1961, double A units	50	100			
	(B) 1962, AB units	75	140			
230	C & O Alco A unit, 1961	50	100			
231	Alco A unit, 1961-63	50	125			
232	New Haven Alco A unit, 1962	50	90			
233	Steam 2-4-2, 233W Tender, 1961-62	12	25			
235	Steam 2-4-2, 1961	15	30			
236	Steam 2-4-2, 1961-62	10	25			
237	Steam 2-4-2, 1963-66	10	25			
238	Steam 2-4-2, 243W Tender, 1963-64	10	35			
239	Steam 2-4-2, 1965-66	10	35			
240	Steam 2-4-2, circa 1964	15	85			
241	Steam 2-4-2, 1958	20	40			
242	Steam 2-4-2, 1962-66	15	25			
243	Steam 2-4-2, 243W Tender, 1960	15	25			
244	Steam 2-4-2, 244T/1130T Tdr., 1960-61	15	25			
245	Steam 2-4-2, 1959	15	25			
246	Steam 2-4-2, 244T/1130T Tdr., 1959-61	15	30			
247	Steam 2-4-2, 247T Tender, 1959-61	20	40			
248	Steam 2-4-2, 1958	15	40			
249	Steam 2-4-2, 250T Tender, 1958	15	40			
250	Steam 2-4-2, 250T Tender, 1957	15	40			
251	Steam 2-4-2, 250T Tender	15	30			
252	Crossing Gate, 1950-62	7	20			
253	Block Control Signal, 1956-59	10	30			
256	Operating Freight Station, 1950-53	6	20			
257	Freight Station w/diesel horn, 1956-57	20	65			
260	Bumper, die-cast, 1949-69	3	10			
262	Highway Crossing Gate, 1962-69	6	20			
264	Oper. Fork Lift Platform, 1957-60	50	140			

Postwar	Good	Exc	Color	Cond	$
270 Metal Bridge (O)	15	30	___	___	___
282 Gantry Crane, 1954	60	150	___	___	___
282R Gantry Crane, 1956	60	150	___	___	___
299 Code Transmitter Beacon Set, 1961-63	20	55	___	___	___
308 Railroad Sign Set, die-cast, 1945-49	10	25	___	___	___
309 Yard Sign Set, die-cast, 1950-59	3	12	___	___	___
310 Billboard, 1950-68	1	2	___	___	___
311 Billboard, 1969	1	2	___	___	___
313 Bascule Bridge, 1946-49	150	300	___	___	___
313-82 Fiber Pins, 1946-60	.05	.05	___	___	___
313-121 Fiber Pins, 1961	—	1.50	___	___	___
314 Scale Model Girder Bridge, 1945-50	2	10	___	___	___
315 Trestle Bridge, 1946-47	15	45	___	___	___
316 Trestle Bridge, 1949	10	30	___	___	___
317 Trestle Bridge, 1950-56	5	15	___	___	___
321 Trestle Bridge, 1958-64	5	10	___	___	___
332 Arch-Under Bridge, 1959-66	10	30	___	___	___
334 Operating Dispatching Board, 1957-60	50	125	___	___	___
(334) Missile Launching Car, olive drab, 1963-64	60	275	___	___	___
342 Culvert Loader, 1956-58	50	125	___	___	___
345 Culvert Unloading Station, 1958-59	70	200	___	___	___
346 Culvert Unloader, 1965-67	50	125	___	___	___
347 Cannon Firing Range Set, 1962-64	75	150	___	___	___
348 Culvert Unloader, 1966-69	50	125	___	___	___
350 Engine Transfer Table, 1957-60	65	175	___	___	___
350-50 Transfer Table Extension, 1957-60	25	90	___	___	___
352 Ice Depot, 1955-57	50	125	___	___	___
353 Trackside Control Signal, 1960-61	5	20	___	___	___
356 Operating Freight Station, 1952-57	30	70	___	___	___
362 Barrel Loader, 1952-57	30	60	___	___	___
362-78 Wooden Barrels, 1952-57	2	6	___	___	___
364 Conveyor Lumber Loader, 1948-67	30	60	___	___	___
364C On/Off Switch, 1959-64	1	2	___	___	___
365 Dispatching Station, 1958-59	35	75	___	___	___
375 Turntable, 1962-64	50	145	___	___	___
390C Switch, d.p.d.t., 1960-64	.50	2	___	___	___
394 Rotary Beacon, 1949-53	8	20	___	___	___
395 Floodlight Tower, 1949-56	8	30	___	___	___
397 Diesel Operating Coal Loader, 1948-57	30	75	___	___	___
400 B & O RDC Passenger Car, 1956-58	100	200	___	___	___
404 B & O RDC Baggage-mail, 1957-58	150	350	___	___	___
410 Billboard Blinker, 1956-58	10	30	___	___	___
413 Countdown Control Panel, 1962	6	12	___	___	___
415 Diesel Fueling Station, 1955-57	30	100	___	___	___
419 Heliport Control Tower, 1962	75	200	___	___	___
443 Missile Launch Platform, 1960-62	15	20	___	___	___
445 Switch Tower, lighted, 1952-57	20	50	___	___	___
448 Missile Firing Range Set, 1961-63	25	50	___	___	___
450 Signal Bridge, two-track, 1952-58	15	35	___	___	___
452 Signal Bridge, single-track, 1961-63	40	80	___	___	___
455 Operating Oil Derrick, 1950-54	50	140	___	___	___
456 Coal Ramp/ 3456 Hopper, 1950-55	100	180	___	___	___
460 Piggyback Transportation, 1955-67	20	50	___	___	___

Postwar		Good	Exc	Color	Cond	$
460P	Piggyback Platform	20	40			
461	Platform with Truck and Trailer	25	70			
462	Derrick Platform Set, 1961-62	30	90			
464	Lumber Mill, 1956-60	30	90			
465	Sound Dispatching Station, 1956-57	30	60			
470	Missile Launching Platform, 1959-62	12	25			
480-25	Conversion Coupler, 1950-60	.50	1.50			
480-32	Conv. Magnetic Coupler, 1961-69	.50	1.50			
494	Rotary Beacon, 1954-66	10	30			
497	Coaling Station, no car, 1953-58	50	125			
520	Lionel Lines Switcher, 1956-57	35	90			
600	M K T NW-2 Switcher, 1955	75	150			
601	Seaboard NW-2 Switcher, 1956	75	150			
602	Seaboard NW-2 Switcher, 1957-58	90	175			
610	Erie NW-2 Switcher, 1955	75	150			
611	Jersey Central NW-2, 1957-58	75	150			
613	Union Pacific NW-2 Switcher, 1958	75	200			
614	Alaska NW-2 Switcher, 1959-60	90	160			
616	Santa Fe NW-2, 1961-62	75	175			
617	Santa Fe NW-2 Switcher, 1963	100	200			
621	Jersey Central NW-2, 1956-57	50	125			
622	Santa Fe NW-2 Switcher, 1949-50	100	275			
623	Santa Fe NW-2 Switcher, 1952-54	75	200			
624	C & O NW-2 Switcher, 1952-54	125	250			
625	L V GE 44-ton Switcher, 1957-58	50	130			
626	B & O GE 44-ton Switcher, 1959	175	275			
627	L V GE 44-ton Switcher, 1956-57	75	125			
628	N P GE 44-ton Switcher, 1956-57	75	125			
629	Burlington GE 44-ton, 1956	75	250			
633	Santa Fe NW-2 Switcher, 1962	50	100			
634	Santa Fe NW-2, 1963, 1965-66	40	80			
635	Union Pacific NW-2 Switcher, 1965	40	90			
637	Steam 2-6-4, 2046W Tender, 1959-63	40	85			
638-2361	Van Camps Pork & Beans Box., 62	10	20			
645	Union Pacific NW-2 Switcher, 1969	40	90			
646	Steam 4-6-4, 2046W Tender, 1954-58	75	175			
665	Steam 4-6-4, 2046W/6026W Tdr., 54-59	60	135			
670	Pennsylvania turbine, 6-8-6, 1952			Not Manufactured		
671R	Steam 6-8-6, 4424W/4671 Tdr., 46-49	100	250			
671	Steam 6-8-6, 671W Tender, 1946-49	50	150			
671RR	Steam 6-8-6, 671W Tender, 1952	75	150			
674	Steam 2-6-4, 1952			Not Manufactured		
675	Steam 2-6-2, 2466W/2466WX/ 6466WX Tender, 1947	45	115			
681	Steam Turbine, 6-8-6, 2046W-50/2671W/ 2671WX Tender, 1950-51, 1953	75	170			
682	Steam 6-8-6, 2046W-50 Tdr., 1954-55	150	250			
685	Steam 4-6-4, 6026W Tender, 1953	75	175			
686	Steam 4-6-4, circa 1953-54	NRS	NRS			
703	Steam 4-6-4, Hudson, 1946			Not Manufactured		
703-10	Special Smoke Bulb, 1946	—	3			
725	Steam 2-8-4, Berkshire, 1952			Not Manufactured		
726	(A) Steam 2-8-4 Berkshire, 1946	200	375			
726	(B) (C) Steam 2-8-4, 1947-49	200	325			

	Good	Exc	Color	Cond	$
726RR Steam 2-8-4 Berkshire, 1952	100	225	___	___	___
736 2-8-4, 2046W/736W/267W, 1950-66	175	275	___	___	___
746 (A) Steam 4-8-4, N & W, long stripe Tender., 1957	500	1450	___	___	___
746 (B) Steam 4-8-4, N & W, short stripe Tender, 1958-60	400	1200	___	___	___
760 Curved Track, 16 sections (O72), 54-57	15	35	___	___	___
773 (A) Steam 4-6-4, 2426W Tdr., 1950	650	1850	___	___	___
773 (B) Steam 4-6-4, 1964	475	1350	___	___	___
773 (C) Steam 4-6-4, 1965-66	475	1375	___	___	___
909 Smoke Fluid, 1957-58	—	2	___	___	___
919 Artificial Grass, 1946-64	—	3	___	___	___
1920 Scenic Display Set, 1957-58	15	60	___	___	___
920-2 Tunnel Portals, pair, 1958	8	20	___	___	___
920-5 Artificial Rock, 1958	.50	2	___	___	___
920-8 Lichen, 1958	.50	2	___	___	___
925 Lionel Lubricant, tube, 1946-69	.50	2	___	___	___
926 Lionel Lubricant, tube, 1955	.25	1	___	___	___
926-5 Instruction Booklet, 1946-48	.25	1	___	___	___
927 Lubricating Kit, 1950-53	2	7	___	___	___
928 Maintenance & Lubricating Kit, 1960-63	2	5	___	___	___
943 Ammo Dump, 1959-61	4	8	___	___	___
950 U. S. Railroad Map, 1958-66	3	12	___	___	___
951 Farm Set, 1958	3	20	___	___	___
952 Miniature Figure Set, 1958	3	15	___	___	___
953 Miniature Figure Set, 1960-62	2	15	___	___	___
954 Swimming Pool & Playground Set, 1959	3	15	___	___	___
955 Farm Building and Animal Set, 1958	3	15	___	___	___
956 Stockyard Set, 1959	3	15	___	___	___
957 Farm Building and Animal Set, 1958	3	15	___	___	___
958 Vehicle Set, 1958	3	15	___	___	___
959 Barn Set, 1958	3	15	___	___	___
960 Barnyard Set, 1959-61	3	15	___	___	___
961 School Set, 1959	3	15	___	___	___
962 Turnpike Set, 1958	3	18	___	___	___
963 Frontier Set, 1959-60	3	15	___	___	___
964 Factory Set, 1959	3	18	___	___	___
965 Farm Set, 1959	3	15	___	___	___
966 Fire House Set, 1958	3	15	___	___	___
967 Post Office Set, 1958	3	15	___	___	___
968 TV Transmitter Set, 1958	7	20	___	___	___
969 Construction Set, 1960	10	30	___	___	___
970 Ticket Booth, 1958-60	25	50	___	___	___
971 Lichen Package, 1960-64	2	4	___	___	___
972 Landscape Tree Assortment, 1961-64	2	4	___	___	___
973 Complete Landscaping Set, 1960-64	3	6	___	___	___
974 Scenery Set, 1962-63	4	8	___	___	___
980 Ranch Set, 1960	3	15	___	___	___
981 Freight Yard Set, 1960	3	15	___	___	___
982 Suburban Split Level Set, 1960	3	15	___	___	___
983 Farm Set, 1960-61	10	35	___	___	___
984 Railroad Set, 1961-62	3	15	___	___	___
985 Freight Area Set, 1961	7	18	___	___	___
986 Farm Set, 1962	5	15	___	___	___

		Good	Exc	Color	Cond	$
987	Town Set, 1962	10	35			
988	Railroad Structure Set, 1962	6	15			
1001	Steam 2-4-2, 1001T Tender, 1948	12	25			
1002	Lionel Gon., silver, yellow, or red, 1949	50	195			
1002	Lionel Gondola, blue or black, 1949	2	5			
X1004	P R R Baby Ruth Boxcar, 1948-51	2	4			
1005	Sunoco 1-D Tank Car, 1948-50	2	5			
1007	Lionel Lines Cab., SP Type 3, 1948-52	2	5			
1008	Uncoupling Unit (O27), 1957	.50	1			
1008-50	Uncoupling Track (O27), 1948	.25	1			
1010	Transformer, 35 watts, 1961-66	3	5			
1011	Transformer, 25 watts, 1948-49	3	5			
1012	Transformer, 40 watts, 1950-54	3	5			
1013	Curved Track (O27), 1945-69	.10	.20			
1013-17	Steel Pins (O27), 1946-60	—	.05			
1013-42	Steel Pins (O27), 1961-68	—	.60			
1015	Transformer, 45 watts, 1956-60	5	8			
1016	Transformer, 35 watts, 1959-60	3	5			
1018-1/2	Straight Track (O27), 1955-59	.10	.30			
1018	Straight Track (O27), 1945-69	.10	.30			
1019	Remote Control Track Set (O27)	1.50	5			
1020	90° Crossing (O27), 1955-69	1.50	3			
1021	90° Crossing (O27), 1945-54	1.50	3			
1022	Manual Switches, pair (O27), 1953-69	2	10			
1023	45° Crossing (O27), 1955-69	1.50	3			
1024	Manual Switches, pair (O27), 1946-52	5	10			
1025	Transformer, 45 watts, 1961-69	3	10			
1025	Illuminated Bumper (O27), 1946-47	1	1.50			
1026	Transformer, 25 watts, 1963-64	1	1.50			
1032	Transformer, 75 watts, 1948	9	14			
1033	Transformer, 90 watts, 1948-56	20	30			
1034	Transformer, 75 watts, 1948-54	15	20			
1037	Transformer, 40 watts, 1946-47	1	1.50			
1041	Transformer, 60 watts, 1945-46	5	7			
1042	Transformer, 75 watts, 1947-48	9	14			
1043	(A) Trans., 50 watts, black, 1953-57	2	3			
1043	(B) Transformer, 60 watts, ivory, 1957	40	50			
1044	Transformer, 90 watts, 1957-59	20	30			
1045	Operating Watchman, 1946-50	8	20			
1047	Operating Switchman, 1959-61	30	100			
1050	Steam 0-4-0, 1050 Tender (O27), 1959	12	25			
1053	Transformer, 60 watts, 1956-60	5	8			
1055	Texas Special Alco A unit, 1959-60	25	75			
1060	Steam 2-4-2, 1050T, 1130T Tdr., 60-61	12	25			
1061	Steam 0-4-0, 1961-64; 2-4-2, 1969	12	25			
1062	Steam 2-4-2, 1963-64	12	25			
1063	Transformer, 75 watts, 1960-64	9	14			
1065	Union Pacific Alco A unit, 1961	25	75			
1066	Union Pacific Alco A unit, 1964	25	75			
1073	Transformer, 60 watts, 1962-66	3	5			
1101	Steam 2-4-2, 1948	12	25			
1101	Transformer, 25 watts, 1948	1	1.50			
1110	Steam 2-4-2, 1949, 1951-52	12	25			
1120	Steam 2-4-2, 1950	12	25			

		Good	Exc	Color	Cond	$
1121	R.C. Switches, pair (O27), 1946-51	7	25	___	___	___
1122	R.C. Switches, pair (O27), 1952-53	10	30	___	___	___
1122-34	R.C. Switches, pair, 1952-53	10	35	___	___	___
1122-500	Gauge Adapter (O27), 1957-66	.25	1	___	___	___
1122E	R.C. Switches, pair (O27), 1953-69	10	35	___	___	___
1130	Steam 2-4-2, 1130T Tender, 1953-54	12	25	___	___	___
1615	Steam 0-4-0, 1615T Tender, 1955-57	55	140	___	___	___
1625	Steam 0-4-0, 1625 Tender, 1958	65	200	___	___	___
1640-100	Presidential Kit, 1960	10	25	___	___	___
1654	Steam 2-4-2, 1654T/1654W Tdr., 46-47	15	30	___	___	___
1655	Steam 2-4-2, 6654W Tender, 1948-49	15	35	___	___	___
1656	Steam 0-4-0, 6403 Tender, 1948-49	125	225	___	___	___
1665	Steam 0-4-0, 2403B Tender, 1946	125	275	___	___	___
1666	Steam 2-6-2, 6654W/2466WX, 46-47	25	70	___	___	___
1862	General 4-4-0, 1862 Tender, 1959-62	75	175	___	___	___
1865	Western & Atlantic Coach, 1959-62	13	25	___	___	___
1866	Western & Atlantic Baggage, 1959-62	15	35	___	___	___
1872	General 4-4-0, 1872T/1875W Tender, 1959-62	100	250	___	___	___
1875	Western & Atlantic Coach, 1959-62	40	100	___	___	___
1875W	Western & Atlantic Coach, 1959-62	40	80	___	___	___
1876	Western & Atlantic Baggage, 1959-62	25	60	___	___	___
1877	No number, flatcar, 1960-65	3	5	___	___	___
1877	Flatcar w/fence and horses, 1959-62	20	50	___	___	___
1882	General 4-4-0, 1882T Tdr., 1959-62	150	400	___	___	___
1885	Western & Atlantic Coach, 1959	75	250	___	___	___
1887	Flatcar w/fences and horses, 1959	50	225	___	___	___
2003	Track "Make-up" Kit for "O27 Track", 1963	NRS	NRS	___	___	___
2016	Steam 2-6-4, 6026W Tender, 1955-56	20	60	___	___	___
2018	Steam 2-6-4, 6026W Tdr., 1956-59, 61	25	65	___	___	___
2020	Steam 6-8-6, 2020W Tender, 1946-47	75	140	___	___	___
2023	Union Pacific Alco AA units, 1950-51	100	250	___	___	___
2024	Chesapeake & Ohio Alco A, 1969	35	75	___	___	___
2025	Steam 2-6-2, 2-6-4, 2460WX/2466WX/6466WX/6466W Tender 1947-49, 1952	50	95	___	___	___
2026	Steam 2-6-2, 6466T/6466W/6466WX Tender, 1948-49, 1951-53	40	70	___	___	___
2028	Pennsylvania GP-7, 1955	125	350	___	___	___
2029	Steam 2-6-4, 243W Tender, 1964-69	40	70	___	___	___
2031	Rock Island Alco AA, 1952-54	125	400	___	___	___
2032	Erie Alco AA units, 1952-54	100	225	___	___	___
2033	Union Pacific Alco AA, 1952-54	100	250	___	___	___
2034	Steam 2-4-2, 1952	20	35	___	___	___
2035	Steam 2-6-4, 2466W Tender, 1950-51	50	75	___	___	___
2036	Steam 2-6-4, 6466W Tender, 1950	40	70	___	___	___
2037	Steam 2-6-4, black engine 1954-55, 1957-58, 1962-63	50	85	___	___	___
2037-500	Steam 2-6-4, pink engine, 1957*	300	500	___	___	___
2041	Rock Island Alco AA, 1969	50	110	___	___	___
2046	Steam 4-6-4, 2046W Tdr., 1950-51, 53	100	175	___	___	___
2055	Steam 4-6-4, 1025W/2046W, 1953-55	65	140	___	___	___
2056	Steam 4-6-4, 2046W Tender, 1952	90	190	___	___	___
2065	Steam 4-6-4, 2046W/6026W, 1954-57	80	140	___	___	___

		Good	Exc	Color	Cond	$
2240	Wabash F-3 AB units, 1956	300	700			
2242	New Haven F-3 AB units, 1958-59	400	850			
2243	Santa Fe F-3 AB units, 1955-57	225	400			
2243C	Santa Fe F-3 B unit, 1955-57	75	175			
2245	Texas Special F-3 AB, 1954-55					
(A)	1954, cabs w/portholes, 2-pc. horn	225	500			
(B)	1955, cabs w/o portholes, 1-pc. horn	375	700			
2257	Lionel SP Type 1 caboose, 1947	3	6			
2321	Lackawanna Trainmaster, 1954-56					
(A)	Gray roof	200	475			
(B)	Maroon roof	300	700			
2322	Virginian Trainmaster, 1965-66	325	575			
2328	Burlington GP-7, 1955-56	150	375			
2329	Virginian Rectifier, 1958-59	250	650			
2330	Pennsylvania GG-1, green, 1950	275	1250			
2331	Virginian Trainmaster, 1955-58					
(A)	1955, black stripe & gold lettering	550	1250			
(B)	1956-58, blue stripe & yellow lett.	350	750			
2332	Pennsylvania GG-1, 1947-49	550	1250			
2333	Santa Fe F-3 AA units, 1948-49	225	500			
2333	N Y C F-3 AA units, 1948-49	225	500			
2337	Wabash GP-7, 1958	100	300			
2338	Milwaukee Road GP-7, 1955-56					
(A)	Orange band all around shell	750	1750			
(B)	Incomplete orange band	100	225			
2339	Wabash GP-7, 1957	100	250			
2340	Pennsylvania GG-1, 1955	350	1500			
2341	Jersey Central Trainmaster, 1956					
(A)	High gloss orange	1000	2250			
(B)	Dull orange	800	1950			
2343	Santa Fe F-3 AA units, 1950-52	200	650			
2343C	Santa Fe F-3 B unit, 1950-55	65	195			
2344	N Y C F-3 AA units, 1950-52	175	575			
2344C	N Y C F-3 B unit, 1950-55	100	250			
2345	Western Pacific F-3 AA, 1952	800	1800			
2346	Boston & Maine GP-9, 1965-67	100	250			
2347	C & O GP-7, 1965	1250	2750			
2348	M & St L GP-9, 1958-59	150	425			
2349	Northern Pacific GP-9, 1959-60	200	375			
2350	New Haven EP-5, 1956-58					
(A)	White "N" & "NEW HAVEN", orange "H"	350	750			
(B)	Orange "N" & "NEW HAVEN", black "H"	950	1500			
2351	Milwaukee EP-5, 1957-58	200	500			
2352	Pennsylvania EP-5, 1958-59	225	525			
2353	Santa Fe F-3 AA units, 1953-55	225	600			
2354	N Y C F-3 AA units, 1953-55	225	525			
2355	Western Pacific F-3 AA, 1953	600	1550			
2356	Southern F-3 AA units, 1954-56	450	1250			
2356C	Southern F-3 B unit, 1954-56	125	325			
2357	Lionel SP Type 1 or 2 Caboose, 47-48	10	20			
2358	Great Northern EP-5, 1959-60	300	750			
2359	Boston & Maine GP-9, 1961-62	100	250			

		Good	Exc	Color	Cond	$
2360	Penn GG-1, 1956-58, 1961-63					
(A)	Tuscan, five gold stripes	475	1450	___	___	___
(B)	Dark green, five gold stripes	475	1150	___	___	___
(C)	Tuscan, single gold stripe	350	850	___	___	___
2363	Illinois Central F-3 AB, 1955-56	300	1200	___	___	___
2365	C & O GP-7, 1962-63	125	350	___	___	___
2367	Wabash F-3 AB units, 1955	325	1000	___	___	___
2368	B & O F-3 AB units, 1956	700	2200	___	___	___
2373	Canadian Pacific F-3 AA units, 1957	700	1800	___	___	___
2378	Milwaukee Road F-3 AB units, 1956	700	1950	___	___	___
2379	Rio Grande F-3 AB units, 1957-58	400	1100	___	___	___
2383	Santa Fe F-3 AA units, 1958-66	125	225	___	___	___
2400	Maplewood Pullman, green, 1948-49	25	100	___	___	___
2401	Hillside Observation, green, 1948-49	25	100	___	___	___
2402	Chatham Pullman, green, 1948-49	25	100	___	___	___
2404	Santa Fe Vista Dome, 1964-65	11	30	___	___	___
2405	Santa Fe Pullman, 1964-65	12	30	___	___	___
2406	Santa Fe Observation, 1964-65	12	30	___	___	___
2408	Santa Fe Vista Dome, 1966	12	35	___	___	___
2409	Santa Fe Pullman, 1966	15	35	___	___	___
2410	Santa Fe Observation, 1966	15	35	___	___	___
2411	Flatcar w/pipes, die-cast, 1946	20	100	___	___	___
2411	Flatcar w/logs, die-cast, 1947-48	10	25	___	___	___
2412	Santa Fe Vista Dome, 1959-63	12	40	___	___	___
2414	Santa Fe Pullman, 1959-63	12	40	___	___	___
2416	Santa Fe Observation, 1959-63	12	40	___	___	___
2419	D L & W Work Caboose, 1946-47	15	30	___	___	___
2420	D L & W Work Cab., w/light, 1946-48	25	80	___	___	___
2421	Maplewood Pullman, 1950- 53	20	40	___	___	___
2422	Chatham Pullman, 1950-53	20	40	___	___	___
2423	Hillside Observation, 1950-53	20	40	___	___	___
2429	Livingston Pullman, 1952-53	20	40	___	___	___
2430	Pullman, 1946-47	12	25	___	___	___
2431	Observation, 1946-47	20	40	___	___	___
2432	Clifton Vista Dome, 1954-58	15	35	___	___	___
2434	Newark Pullman, 1954-58	15	35	___	___	___
2435	Elizabeth Pullman, 1954-58	15	35	___	___	___
2436	Summit Observation, 1954-58	20	40	___	___	___
2436	Mooseheart Observation, 1957-58	30	65	___	___	___
2440	Pullman, 1946-47	20	40	___	___	___
2441	Observation, 1946-47	20	40	___	___	___
2442	Clifton Vista Dome, 1956	20	55	___	___	___
2442	Pullman 1946-47	20	55	___	___	___
2443	Observation, 1946-47	20	55	___	___	___
2444	Newark Pullman, 1956	20	55	___	___	___
2445	Elizabeth Pullman, 1956	25	65	___	___	___
2446	Summit Observation, 1956	20	55	___	___	___
2452	Pennsylvania Gondola, 1945-47	5	15	___	___	___
2452X	Pennsylvania Gondola, 1946-47	4	10	___	___	___
X2454	Pennsylvania Boxcar, 1946	20	100	___	___	___
X2454	Baby Ruth Box., "P R R" logo, 1946-47	5	15	___	___	___
2456	Lehigh Valley Short Hopper, 1948	5	15	___	___	___
2457	Penn. Caboose, metal, N5, 1945-47	10	25	___	___	___
X2458	Pennsylvania Boxcar, 1946-47	15	35	___	___	___

		Good	Exc	Color	Cond	$
2460	Bucyrus Erie Crane, 12-wheel, 1946-50					
	(A) Black letters	60	200			
	(B) White letters	35	75			
2461	Transformer Car, die-cast, 1947-48	30	80			
2465	Sunoco 2-D Tank Car, 1946-48	5	15			
2472	Penn. Caboose, metal, N5, 1946-47	8	20			
2481	Plainfield Pullman, yellow, 1950	50	160			
2482	Westfield Pullman, yellow, 1950	50	160			
2483	Livingston Observation, yellow, 1950	40	130			
2521	President McKinley Obs., 1962-66	40	80			
2522	President Harrison V. D., 1962-66	40	80			
2523	President Garfield Pullman, 1962-66	48	80			
2530	Railway Exp. Agency Bagg., 1954-60	50	100			
2531	Silver Dawn Observation, 1952-60	25	60			
2532	Silver Range Vista Dome, 1952-60	25	60			
2533	Silver Cloud Pullman, 1952-60	25	60			
2534	Silver Bluff Pullman, 1952-60	25	60			
2541	Alexander Hamilton Obs., 1955-56*	70	190			
2542	Betsy Ross Vista Dome, 1955-56*	70	190			
2543	William Penn Pullman, 1955-56*	70	190			
2544	Molly Pitcher Pullman, 1955-56*	70	190			
2550	B & O RDC Bagg. Mail Car, 1957-58	250	550			
2551	Banff Park Observation, 1957*	95	225			
2552	Skyline 500 Vista Dome, 1957*	95	225			
2553	Blair Manor Pullman, 1957*	150	325			
2554	Craig Manor Pullman, 1957*	150	325			
2555	Sunoco 1-D Tank Car, 1946-48	12	35			
2559	B & O RDC Passenger Car, 1957-58	200	325			
2560	Lionel Lines Crane, 8-wheel, 1946-47	20	60			
2561	Vista Valley Observation, 1959-61*	95	225			
2562	Regal Pass Vista Dome, 1959-61*	125	300			
2563	Indian Falls Pullman, 1959-61*	100	300			
2625	Madison Pullman, 1946-47*	75	250			
2625	Manhattan Pullman, 12-wheel, 46-47*	75	250			
2625	Irvington Pullman, 12-wheel, 1946-50*	75	225			
2627	Madison Pullman, 12-wheel, 1948-50*	75	225			
2628	Manhattan Pullman, 12-wheel, 48-50*	75	225			
2630	Pullman, metal, 1938-42	NRS	NRS			
2631	Observation, metal, 1946	NRS	NRS			
2671	T C A Tender, 1968 (Like New price)	—	75			
2755	S U N X 1-D Tank Car, 1945	40	150			
2855	S U N X 1-D Tank Car, 1946-47	50	150			
2856	B & O Scale Hopper Car, 1946-47			Not Manufactured		
2857	N Y C Scale Caboose, 1946			Not Manufactured		
X2954	Penn Scale Boxcar, 1941-42	150	300			
2955	S U N X 1-D Scale Tank, 1940-42, 46	110	275			
2956	B & O Scale Hopper Car, 1940-42	150	325			
2957	N Y C Scale Caboose, 1946	70	250			
(3309)	Turbo Missile Launch Car, 1963-64	20	50			
3330	Flatcar with Submarine kit, 1960-61	35	120			
3330-100	Operating Submarine Kit, 1960-61	10	25			
(3349)	Missile Launching Car, red, 1962-65	20	50			
3356	Santa Fe Express, 1956-60, 1964-66	20	45			
3356-150	Horse Corral Car Set	10	25			

	Good	Exc	Color	Cond	$
3357 Hydraulic Maintenance Car	15	45			
3359-55 Lionel Lines Two-bin Dump, 55-58	10	35			
3360 Operating Burro Crane, 1956-57	150	300			
3361-55 Oper. Log Dump Car, 1955-58	10	25			
3362 Flatcar with Helium Tanks, 1961-63	10	35			
3364 Log Dump Car, 1965-69	15	45			
3366 Circus Car Corral Set, 1959-62	90	175			
3366 Circus Car, 1959-62	45	70			
3366-100 (9) White Horses, 1959-60	5	10			
3370 Western & Atlantic Outlaw Car, 61-64	15	40			
3376 Bronx Zoo Car, 1960-69	15	40			
3376 Giraffe Activator Unit	12	25			
3386 Bronx Zoo Car (O27), 1960	20	60			
3409 Helicopter Car, 1961-62	15	100			
3409 Satellite Car, circa 1961	35	100			
3410 Helicopter Car, 1961-62	20	50			
3413 Mercury Capsule Car, 1962-64	40	110			
3419 Helicopter Car, 1959-65	25	90			
3424-100 Low Bridge Signal Set	10	20			
3424 Wabash Operating Boxcar, 1956-58	25	60			
3428 U. S. Mail Operating Boxcar, 1959	25	75			
3429 U S M C Helicopter Car, 1960					
(A) Navy helicopter	45	120			
(B) U S M C helicopter	100	375			
3434 Poultry Dispatch Car, 1959-60, 65-66	30	90			
3435 Traveling Aquarium Car, 1959-62					
(A) With "Tank No. 1" and "Tank No. 2"	100	500			
(B) Without tank designations	40	85			
3444 Erie Operating Gondola, 1957-59	25	50			
3449 Lionel Lines Ore Dump Car, 1946	NRS	NRS			
3451 Operating Log Dump Car, 1946-48	10	40			
3454 P R R Oper. Merchandise Car, 1946-47	35	100			
3456 N & W Operating Hopper, 1950-55	15	40			
3459 Lionel Lines Oper. Dump Car, 1946-48	20	50			
3460 Flatcar with trailers, 1955	20	45			
3461 Lionel Operating Log Car, 1949-55	10	40			
3462 Automatic Milk Car, 1947-48	30	75			
3462P Milk Car Platform	2	5			
X3464(A) A T & S F Oper. Boxcar, 1949-53	10	25			
X3464(B) N Y C Operating Boxcar, 1949-52	10	25			
3469 Lionel Lines Oper. Dump Car, 1949-55	10	35			
3470 Target Launcher, 1962-64	20	60			
3472 Automatic Milk Car, 1953	20	40			
3474 Western Pacific Boxcar, 1952-53	15	50			
3482 Automatic Milk Car, 1954-55	20	60			
3484 Pennsylvania Operating Boxcar, 1953	15	50			
3484-25 A T & S F Oper. Boxcar, 1954	20	90			
3494-1 N Y C Pacemaker Boxcar, 1955	20	90			
3494-150 M P Operating Boxcar, 1956	45	125			
3494-275 State of Maine Oper. Box., 1956-58	40	75			
3494-550 Monon Operating Box., 1957-58	100	275			
3494-625 SOO Operating Boxcar, 1957-58	100	275			
3509 Satellite Car, 1961	15	50			

	Good	Exc	Color	Cond	$
3510 Satellite Car, 1961-62					
(A) Gray superstructure	30	125	___	___	___
(B) No superstructure	10	30	___	___	___
3512 Ladder Co. Car, 1959-61					
(A) Black rooftop ladder	25	70	___	___	___
(B) Silver rooftop ladder	35	125	___	___	___
3519 Satellite Car, 1961-64	15	50	___	___	___
3520 Searchlight Car, die-cast, 1952-53	20	50	___	___	___
3530 G M Electro Mobile Power, 1956-58	40	70	___	___	___
3530 Searchlight Car with pole and base	25	75	___	___	___
3535 A E C Security Car, 1960-61	20	90	___	___	___
3540 Operating Radar Car, 1959-60	40	125	___	___	___
3545 Lionel TV Car, 1961-62	40	125	___	___	___
3559 Operating Coal Dump Car, 1946-48	10	20	___	___	___
3562-1 A T & S F Oper. Barrel Car, 1954	60	150	___	___	___
3562-25 A T & S F Operating Barrel Car (red lettering), 1954	100	300	___	___	___
3562-25 A T & S F Operating Barrel Car (blue lettering), 1954	15	30	___	___	___
3562-50 A T & S F Oper. Barrel Car, 55-57	15	40	___	___	___
3562-75 A T & S F Oper. Barrel Car, 1958	20	50	___	___	___
3619 Helicopter Car, 1962-64	30	75	___	___	___
3620 Searchlight Car, 1954-56	25	50	___	___	___
3620X Lionel Lines Searchlight Car, 1955	25	40	___	___	___
3650 Extension Searchlight Car, 1956-59	25	60	___	___	___
3656 Armour Operating Cattle Car, 1949-55	18	50	___	___	___
3656 Stockyard with cattle	18	50	___	___	___
3662-1 Automatic Milk Car, 1955-60, 64-66	30	60	___	___	___
3665 Minuteman Operating Car, 1961-64					
(A) Medium blue roof	75	150	___	___	___
(B) Dark blue roof	20	45	___	___	___
3666 Minuteman Boxcar w/missile, c. 1964	150	500	___	___	___
3672 Bosco Operating Boxcar, 1959-60					
(A) With "Bosco" decal	110	425	___	___	___
(B) Without "Bosco" decal	100	325	___	___	___
3820 Flatcar with Submarine, 1960-62	30	180	___	___	___
3830 Flatcar with Submarine, 1960-63	30	90	___	___	___
3854 Oper. Merchandise Car, 1946-47	200	500	___	___	___
3927 Lionel Lines Track Cleaner, 1956-60	60	110	___	___	___
3927 Track Cleaning Fluid, 1957-69	.50	2	___	___	___
3927-75 Track Cleaning Fluid, 1957-69	.50	2	___	___	___
4357 Penn N5 Caboose, electronic, 1948-50	50	150	___	___	___
4452 Penn Gondola, electronic, 1946-48	30	75	___	___	___
4454 Baby Ruth P R R Box., electr., 1946-47	50	150	___	___	___
4457 Penn N5 Caboose, electr., 1946-47	40	140	___	___	___
4681 Steam 6-8-6, electronic, 1950			Not Manufactured		
4776-18 (See 2472, 2457)					
4810 S P Trainmaster, prototype, 1954	NRS	NRS	___	___	___
5159 Maintenance Kit, 1964-68	1	3	___	___	___
5159-50 Maintenance & Lubricant Kit, 1969	1	3	___	___	___
5160 Viewing Stand	50	100	___	___	___
5364-17 (See 6417)					
5459 L L Dump Car, electronic, 1946-49	30	70	___	___	___
6002 N Y C Gondola (O27), 1950	4	10	___	___	___

Postwar	Good	Exc	Color	Cond	$
6004 Baby Ruth P R R Boxcar (O27), 1950	NRS	NRS			
X6004 Baby Ruth P R R Boxcar (O27) 1950	2	5			
6007 L L SP Type 3 Caboose (O27), 1950	2	5			
6012 Lionel Gondola, 1951-56	2	5			
6014 Airex Boxcar (O27)	40	115			
6014 Bosco P R R Boxcar (O27), 1958					
(A) White with black letters	15	30			
(B) Red or orange body	4	9			
6014 Campbell Soup Boxcar (O27), 1969	NRS	NRS			
6014 Chun King Boxcar (O27)	58	115			
6014 Frisco Boxcar (O27)					
(A) White, Type I body	10	20			
(B) White, other body types	2	6			
(C) Red, orange, or cream	2	6			
X6014 Baby Ruth P R R Boxcar (O27)	2	4			
6014-1 Frisco Boxcar	4	9			
6014-60 Frisco Boxcar, 1957-58	10	25			
6014-85 Frisco Boxcar, 1969	4	9			
6014-100 Frisco Boxcar	4	9			
6014-100 Airex Boxcar, 1960-61	4	9			
6014-150 Wix Boxcar (O27), uncatalogued	50	100			
6014-325 Frisco Boxcar, 1963-64	4	9			
6014-335 Frisco Boxcar, 1965-66	2	6			
6014-410 Frisco Boxcar, 1969	NRS	NRS			
6015 Sunoco 1-D Tank Car (O27), 1954-55	3	5			
6017 Lionel Lines SP Type Caboose (O27)	3	7			
6017-50 U S M C Caboose, 1958	20	50			
6017-100 B & M SP Type Caboose, 1959, 1962, 1965-66					
(A) Purplish blue	150	450			
(B) (C) Medium or light blue	10	30			
6017-185 A T & S F SP Type 4 Caboose, 1959-60	10	30			
6017-200 U. S. Navy SP Type 4 Cab., 1960	25	80			
6017-225 A T & S F Type 4 Cab., c. 1961-62	10	45			
6017-235 A T & S F SP Type 4 Caboose	20	40			
6019 RCS Track Set (O27), 1948-66	1	3			
6024 Nabisco Shredded Wheat Boxcar, 1957	10	25			
6024 R C A Whirlpool Boxcar (O27), 1957	30	55			
6025 Gulf 1-D Tank Car (O27), 1956-57	3	10			
6027 Alaska SP Type 2A Caboose, 1959	25	75			
6029 Uncoupling Track (O27), 1955, 61-63	.25	1			
6032 Lionel Gondola, black (O27), 1952-54	2	5			
X6034 Baby Ruth P R R Boxcar (O27), 1953	5	12			
6035 Sunoco 1-D Tank Car (O27), 1952-53	2	5			
6037 L L Caboose SP Type 3A (O27), 52-54	2	4			
6042 Lionel Gondola (O27), uncat., 1959-62	2	5			
6042-125 Lionel Gondola (O27), blue plastic	2	5			
6044 Airex Boxcar (O27), orange lettering	40	80			
6044 Airex Boxcar, dark blue/purple	70	250			
6044 Airex Boxcar, uncat., yellow lettering	5	12			
6044-1X Boxcar (no letters), c. 1962-63	NRS	NRS			
6045 Lionel Lines 2-D Tank Car, 1958, 1963	3	5			
6045 Cities Service 2-D Tank, uncat., 1960	10	20			
6047 Lionel Lines SP Type 2A/4 Cab., 1962	2	4			

	Good	Exc	Color	Cond	$
6050 Lionel Savings Bank Boxcar, 1961	9	18			
6050 Swift Refrigerator, 1962	10	20			
6050 Libbys Boxcar, no green stem, 1961	50	100			
6050 Libbys Box., green stem on decal, 1961	12	25			
6057 Lionel Lines SP Type 1A/4 Caboose, 1959-62	2	4			
6057-50 Lionel Lines SP Type 4 Cab., 1962	5	25			
6058 C & O SP Type 4 Caboose, 1961	10	45			
6059-50 M St L SP Type 4 Cab., 1961-63	3	8			
6059-60 M St L SP Type 4 Cab., 1963-69	3	8			
6062 N Y C Gondola, 1959-64	3	10			
6062-50 N Y C Gondola, 1969	2	6			
6067 Caboose (no letters), SP Type, 1962	2	4			
6076 Lehigh Valley Hopper, 1963	4	8			
6076 A T S F Hopper, 1963	9	20			
6076 Hopper, gray (no letters)	2	6			
6076-75 Lehigh Valley Hopper Car, 1963	NRS	NRS			
6076-100 Hopper Car (no letters), 1963	2	7			
6109 Flatcar with logs, gray, 1952	35	45			
6110 Steam 2-4-2, 1950-51	12	25			
6111 Flatcar with pipes, red, 1957	5	12			
6111 Flatcar with logs, yellow, 1955-57	3	8			
6111 Flatcar w/unknown load, gray, 1955-57	6	15			
6111-75 Flatcar with logs, 1955, 1958	5	15			
6112 Lionel Gondola, 1956-58	3	8			
(6112) Lionel Gondola, 1963-64	2	5			
6112-1 Lionel Gondola, blue	2	5			
6112-86 Gondola (no letters)	NRS	NRS			
6112-135 Lionel Gondola, black plastic, 1958	2	5			
6119 D L & W Work Caboose, 1955-56	10	25			
6119-25 D L & W Work Caboose, 1956-59	10	25			
6119-50 D L & W Caboose, 1956	10	35			
6119-75 D L & W Caboose, 1957	10	35			
6119-100 D L & W Work Cab., 1957-66, 69	10	20			
6119-125 Caboose (no letters), c. 1960	50	225			
6120 Work Caboose (no letters), 1961-62	7	20			
6121 Flatcar with pipes, yellow, 1955	5	15			
6121-25 Flatcar with logs, yellow, 1955	5	15			
6121-60 Flatcar with pipes, yellow, 1955	5	15			
6130 A T & S F Work Cab., 1961, 65-67, 69	10	30			
6142 Lionel Gondola, 1961-66, 1970	2	5			
6142-50 Lionel Gondola, green, 1961-63, 66	2	5			
6142-75 Lionel Gondola, blue, 1961-63	2	5			
6142-100 Lionel Gondola, green, 1964-65	2	5			
6142-150 Lionel Gondola, blue, c. 1961-63	2	5			
6142-175 Lionel Gondola, c. 1961-63	2	5			
6149 R.C. Uncoupling Track (O27), 1964-69	.25	1			
6151 Flatcar with Patrol Truck, 1958	20	60			
6162 N Y C Gondola, 1961-63	2	7			
6162-60 Alaska Gondola, 1959	20	40			
6162-110 N Y C Gondola, 1964-69	3	7			
6167 Lionel Lines SP Type 4 Caboose, 1963	2	4			
(6167-25) Caboose, red (no letters), SP4, 1964	2	5			
(6167-50) Cab., yellow (no letters), SP4, 1963	2	6			

	Good	Exc	Color	Cond	$
6167-85 U P SP Type 4 Caboose, 1964-69	6	25	___	___	___
(6167-100) Cab. (no letters), SP Type 4, 1964	2	4	___	___	___
6167-125 Cab. (no letters), SP Type 4, 1964	2	4	___	___	___
6167-150 Caboose (no letters), SP Type 4	2	6	___	___	___
6175 Flatcar with rocket, 1958-61	25	60	___	___	___
6176 Lehigh Valley Hopper, 1964-66, 1969	3	7	___	___	___
(6176) Hopper (no letters), yellow	3	7	___	___	___
6219 C & O Work Caboose, 1960	20	65	___	___	___
6220 Santa Fe NW-2 Switcher, 1949-50	100	275	___	___	___
6250 Seaboard NW-2 Switcher, 1954-55	100	275	___	___	___
6257 Lionel SP Type Caboose, 1948-56	2	4	___	___	___
6257X Lionel Lines Caboose	10	25	___	___	___
6257-25 Lionel SP Type 3A Caboose	2	4	___	___	___
6257-50 Lionel SP Type 1 or 3A Caboose	2	4	___	___	___
6257-100 Lionel SP Type 4 Caboose	5	15	___	___	___
6262 Flatcar with wheels, black, 1956-57	15	35	___	___	___
6262 Flatcar with wheels, red, 1956	50	400	___	___	___
6264 Flat. w/lumber for forklift set, 1957-58	20	45	___	___	___
6311 Flatcar, 1955	15	40	___	___	___
6315 Gulf 1-D Chem. Tank, 1956-59, 68-69	10	30	___	___	___
6315 Lionel Lines 1-D Tank Car, 1963-66	7	25	___	___	___
6315-50 Lionel Lines Tank Car	7	25	___	___	___
6315-60 Lionel Lines Tank Car, 1963-64	7	25	___	___	___
6315-1972 T C A Tank Car, 1972	—	75	___	___	___
6342 N Y C Gondola, 1956-58, 1966-69	7	25	___	___	___
6343 Barrel Ramp Car, 1961-62	15	35	___	___	___
6346-56 Alcoa Quad Hopper, 1956	20	45	___	___	___
6352-1 P F E Reefer from Icing Set, 1955-57	25	65	___	___	___
6356-1 N Y C Stock Car, 2 level, 1954-55	10	35	___	___	___
6357 Lionel SP Type Caboose, 1948-61	6	15	___	___	___
6357-25 Lionel SP Type 2 Caboose	6	15	___	___	___
6357-50 A T & S F SP Type 4 Caboose	250	900	___	___	___
6361 Flatcar with timber, 1960-61, 1964-69	25	75	___	___	___
6362-55 Truck Car w/3 trucks, 1955-56	10	30	___	___	___
6376 Lionel Lines Circus Car, 1956-57	20	50	___	___	___
6401 Flatcar with van	3	8	___	___	___
6401-50 Flatcar, gray (no number), c. 1960?	NRS	NRS	___	___	___
6402 Flatcar with reels or boat, 1962, 64, 69	4	12	___	___	___
6402-50 Flatcar with cable reels, 1964	3	10	___	___	___
6404 Flatcar with red auto, 1960	20	60	___	___	___
6404 Flatcar with yellow auto, 1960	—	150	___	___	___
6404 Flatcar with brown or green auto, 1960	—	250	___	___	___
6405 Flatcar with trailer, 1961	10	30	___	___	___
6406 Flatcar with red auto, 1961	15	35	___	___	___
6406 Flatcar with yellow auto, 1961	30	120	___	___	___
6406 Flatcar with brown auto, 1961	—	200	___	___	___
6407 Flatcar with rocket, 1963	75	450	___	___	___
(6408) Flatcar with pipes, 1963	NRS	NRS	___	___	___
6409-25 Flatcar	4	10	___	___	___
6411 Flatcar with logs, 1948-51	10	30	___	___	___
6413 Mercury Project Car, 1962-63	30	150	___	___	___

	Good	Exc	Color	Cond	$
6414 Evans Auto Loader, 1955-57, 1966					
(A) Premium cars w/windows, bumpers, and rubber tires; red, yellow, blue, and white	20	60			
(B) Four cheap cars w/o trim, 2 red, 2 yel.	—	450			
(C) Four red prem. cars w/gray bumpers	50	175			
(D) Four yellow prem. cars w/bumpers	—	450			
(E) Four brown prem. cars w/gray bumpers	—	1700			
(F) Four green prem. cars w/gray bumpers	—	1700			
6415 Sunoco 3-D Tank Car, 1953-55	5	20			
6416 Boat Loader Car, 1961-63	75	150			
6417 Pennsylvania N5C Caboose, 1953-57	10	25			
6417-3 (See 6417-25)					
6417-25 Lionel Lines N5C Caboose, 1954	10	25			
6417-50 L V N5C Caboose, tuscan, 1954	—	1200			
6417-50 Lehigh Valley N5C Cab., gray, 1954	30	100			
6417-53 (See 6417-25)					
6418 Bridge, metal base, plastic sides	1	4			
6418 Flatcar with steel girders, 1955-57	30	65			
6419 D L & W Work Cab., 1948-50, 52-57	10	25			
6419-57 (See 6419-100)					
6419-25 D L & W Work Caboose, 1955	10	30			
6419-50 D L & W Work Caboose, 1956-57	15	35			
6419-75 D L & W Work Caboose, 1956-57	10	30			
6419-100 N & W Work Caboose, 1957-58	40	100			
6420 D L & W Work Caboose, 1948-50	30	85			
6424 Flatcar with two autos, 1956-59	15	35			
6424-85 (See 6424)					
6425 Gulf 3-D Tank Car, 1956-58	10	30			
6426 Sunoco Tank Car, c. 1967	5	20			
6427 Esso Tank Car, c. 1967	5	20			
6427 Lionel Lines N5C Caboose, 1954-60	10	25			
6427-60 Virginian N5C Caboose, 1958	90	300			
6427-500 Pennsylvania N5C Cab., 1957-58*	100	325			
6428 U S Mail Boxcar, 1960-61, 1965-66	10	35			
6429 D L & W Work Caboose, 1963	100	300			
6430 Flatcar w/Cooper-Jarrett vans, 1955-58	15	50			
6431 Flatcar with vans, 1966	40	225			
6434 Poultry Dispatch, 1958-59	30	60			
6436 (See 6436-110)					
6436-1 L V Quad Hopper, black, 1955	10	35			
6436-25 L V Quad Hopper, maroon, 1955-57	10	35			
6436-57 (See 6436-500)					
6436-110 L V Quad Hopper, 1963-68	15	35			
6436-500 L V Hop., lilac, "643657", 57-58*	75	200			
6436-1969 T C A Quad Hopper, 1969	—	125			
6437 Penn N5C Caboose, 1961-68	10	25			
6440 Flatcar with vans, 1961-63	25	75			
6440 Pullman, 1948-49	20	40			
6441 Observation, 1948-49	20	40			
6442 Pullman, 1949	25	60			
6443 Observation, 1949	25	60			
6445 Fort Knox Gold Reserve, 1961-63	40	100			
6446-1 N & W Quad "546446", 1954-55	15	45			
6446-25 N & W Quad "546446", 1955-57	15	35			

		Good	Exc	Color	Cond	$
6446-60	(See 6436-110)					
6447	Pennsylvania N5C Caboose, 1963	80	275	____	____	____
6448	Target Car, 1961-64	10	20	____	____	____
6452	Pennsylvania Gondola, black, 1948-49	4	10	____	____	____
X6454(A)	Baby Ruth P R R Boxcar, 1948	50	200	____	____	____
X6454(B)	N Y C Boxcar, orange, 1948	40	150	____	____	____
X6454(C)	N Y C Boxcar, brown, 1948	20	50	____	____	____
X6454(D)	N Y C Boxcar, tan, 1948	10	30	____	____	____
X6454(E)	A T & S F Boxcar, 1948	10	30	____	____	____
X6454(F)	S P Boxcar, 1949	25	75	____	____	____
X6454(G)	S P Boxcar, 1950	15	40	____	____	____
X6454(H)	S P Boxcar, 1951-52	20	50	____	____	____
X6454(J)	Erie Boxcar, 1949-52	20	50	____	____	____
X6454(K)	Pennsylvania Boxcar, 1949-52	20	50	____	____	____
6456	Lehigh Valley Short Hopper, 1948-55					
	(A) Black	5	9	____	____	____
	(B) Maroon	5	9	____	____	____
6457	Lionel SP Type, 1949-52	10	25	____	____	____
6460	Bucyrus Erie Crane, 8-wheel, 1952-54	15	35	____	____	____
6460-25	Bucyrus Erie Crane, 8-wheel, w/box, 1954	40	80	____	____	____
6461	Transformer Car, 1949-50	25	75	____	____	____
6462	Pennsylvania Gondola, black	2	8	____	____	____
6462	N Y C Gondola, black, 1949-54	2	8	____	____	____
6462	N Y C Gondola, green, 1954, 56-57	3	9	____	____	____
6462	N Y C Gondola, red, 1950-52, 54, 56-57	2	8	____	____	____
6462	Pennsylvania Gondola, black, 1947-48	20	45	____	____	____
6462-25	N Y C Gondola, green	4	10	____	____	____
6462-75	N Y C Gondola, red	4	10	____	____	____
6462-100	N Y C Gondola, red, 1951-54	3	10	____	____	____
6462-125	N Y C Gondola, red, 1954-57	4	9	____	____	____
6462-500	N Y C Gondola, pink, 1957-58*	50	150	____	____	____
6463	Rocket Fuel 2-D Tank, 1962-63	10	25	____	____	____
6464-1	W P Boxcar, blue lettering, 1953-54	25	75	____	____	____
6464-1	W P Boxcar, red lettering, 1953-54	500	1600	____	____	____
6464-25	G N Boxcar, orange, 1953-54	20	50	____	____	____
6464-50	M & St L Boxcar, tuscan, 1953-56	20	50	____	____	____
6464-75	Rock Island Boxcar, 1953-54, 1969	25	75	____	____	____
6464-100	W P Boxcar, 1954-55	30	100	____	____	____
6464-125	N Y C Boxcar, 1954-56	30	100	____	____	____
6464-150	M P Boxcar, 1954-55, 1957	25	110	____	____	____
6464-175	Rock Island Boxcar, 1954-55					
	(A) Blue lettering	25	100	____	____	____
	(B) Black lettering	300	1000	____	____	____
6464-200	P R R Boxcar, 1954-55, 1969	40	125	____	____	____
6464-225	S P Boxcar, 1954-56	35	100	____	____	____
6464-250	W P Boxcar, 1966-67	40	100	____	____	____
6464-275	State of Maine Box., 1955, 57-59	20	75	____	____	____
6464-300	Rutland Boxcar, 1955-56	20	75	____	____	____
6464-325	B & O Sentinel Boxcar, 1956	125	400	____	____	____
6464-350	M K T Katy Boxcar, 1956	80	200	____	____	____
6464-375	Cent. of Ga. Box., 1956-57, 66-67	30	90	____	____	____
6464-400	B & O Timesaver Box., 56-57, 69	20	75	____	____	____
6464-425	New Haven Boxcar, 1956-58	20	40	____	____	____

		Good	Exc	Color	Cond	$
6460-450	G N Boxcar, 1956-57, 1966-67	30	90	___	___	___
6464-475	B & M Boxcar, 1957-59, 66, 68	15	35	___	___	___
6464-500	Timken Boxcar, 1957-59, 1966, 69					
	(A) Yellow & white with charcoal lettering	20	60	___	___	___
	(B) Yellow & white w/red heat-stamped lett.	—	1200	___	___	___
	(C) Green, white, gold, or red heat-stamped lettering	—	750	___	___	___
	(D) "BLT 1-71 BY LIONEL MPC"	20	70	___	___	___
	(E) "9200" on end numberboards and "BLT 1-71 BY LIONEL MPC"	300	1000	___	___	___
6464-510	N Y C Pacemaker Box., 1957-58	150	450	___	___	___
6464-515	M K T Boxcar, 1957-58	150	450	___	___	___
6464-525	M & St L Boxcar	20	50	___	___	___
6464-650	D & R G W Box., 1957-58, 66-67	20	90	___	___	___
6464-700	Santa Fe Boxcar, 1961, 1966-67	20	100	___	___	___
6464-725	New Haven Boxcar, 1962-66, 1968	20	50	___	___	___
6464-825	Alaska Boxcar, yellow and orange lettering, 1959-60	100	150	___	___	___
6464-825	Alaska Box., white letter, 1959-60	—	1200	___	___	___
6464-900	N Y C P & L E Boxcar, 1960-67	25	90	___	___	___
6464-1965	T C A Pittsburgh Boxcar, 1965	—	250	___	___	___
6464-1970	See 1970-89					
6464-1971	See 1970-89					
6465	Sunoco 2-D Tank Car, 1948-56	3	10	___	___	___
6465	Cities Service 2-D Tank Car, 1960-62	6	15	___	___	___
6465	Gulf 2-D Tank Car, black, 1958	25	90	___	___	___
6465	Lionel Lines 2-D Tank Car, 1959, 63-64	5	15	___	___	___
6465-60	Gulf 2-D Tank Car, gray, 1958	10	20	___	___	___
6465-160	(See 6465(D) Lionel Lines)					
6467	Bulkhead Flatcar, 1956	20	50	___	___	___
6468-1	B & O Auto Boxcar, tuscan, 1953-55	100	300	___	___	___
6468-1	B & O Auto Boxcar, blue, 1953-55	20	45	___	___	___
6468-25	New Haven Auto Boxcar, 1956-58	15	65	___	___	___
6469	Lionel Liquified Gases Car, 1963	30	150	___	___	___
6470	Explosives Car, red, 1959-60	10	30	___	___	___
6472	Refrigerator Car, 1950	10	20	___	___	___
6473	Horse Transport Car, 1962-64	10	20	___	___	___
6475	Heinz 57 Vat Car, 1965-66	60	150	___	___	___
6475	Libbys Crushed Pineapple Vat Car, 1963	15	40	___	___	___
6475	Pickles Vat Car	10	25	___	___	___
6476	L V Hopper, 1957-62, 1964, 1968-69	5	9	___	___	___
6476-1	Lehigh Valley Hopper, T T O S, 69	NRS	NRS	___	___	___
6476-125	Lehigh Valley Hopper, 1964	NRS	NRS	___	___	___
6476-135	Lehigh Valley Hopper, 1964-66, 68	5	14	___	___	___
6476-160	Lehigh Valley Hopper, 1969	NRS	NRS	___	___	___
6476-185	Lehigh Valley Hopper, 1969	NRS	NRS	___	___	___
6477	Bulkhead Car with pipes, 1957-58	15	50	___	___	___
6480	Explosives Car, red	10	30	___	___	___
6482	Refrigerator Car, 1957	20	50	___	___	___
6500	Flatcar w/Bonanza plane, 1962, 1965	250	400	___	___	___
6501	Flatcar with jet boat, 1962-63	25	100	___	___	___
6502-50	Flatcar with bridge girder, 1962	15	45	___	___	___
(6502-75)	Flatcar w/bridge girder, 1962	20	60	___	___	___
6511	Flatcar with pipes, 1953-56	10	40	___	___	___

	Good	Exc	Color	Cond	$
6512 Cherry Picker Car, 1962-63	30	80	___	___	___
6517 Lionel Lines B/W Caboose, 1955-59	25	60	___	___	___
6517-75 Erie B/W Caboose, 1966	135	275	___	___	___
6517-1966 T C A B/W Caboose, 1966	60	220	___	___	___
6518 Transformer Car, 1956-58	25	90	___	___	___
6519 Allis Chalmers Flatcar w/load, 1958-61	25	55	___	___	___
6520 Searchlight Car, 1949-51					
(A) Tan diesel generator	150	450	___	___	___
(B)-(D) Green diesel generator	75	225	___	___	___
(E)-(H) Maroon or orange diesel gen.	25	55	___	___	___
6530 Fire Fighting Car, red, 1960-61	25	60	___	___	___
6530 Fire Fighting Car, black, 1960-61	—	750	___	___	___
6536 M & St L Quad Hopper, 1958-59, 1963	15	40	___	___	___
6544 Missile Firing Car, 1960-64					
(A) White-lettered console	25	60	___	___	___
(B) Black-lettered console	100	350	___	___	___
6555 Sunoco 1-D Tank Car, 1949-50	15	45	___	___	___
6556 M K T Cattle Car, 1958	60	200	___	___	___
6557 Lionel SP Type Cab., smoke, 1958-59	75	175	___	___	___
6560 Bucyrus Erie Crane, 8-wheel, 1955-58, 1968-69					
(A) Reddish-orange or black cab, early construction	60	160	___	___	___
(B) Gray cab, lettered frame	40	80	___	___	___
(C) Red cab, unlettered frame	35	110	___	___	___
(D) Red cab, lettered frame	20	50	___	___	___
6560-25 Bucyrus Erie Crane, 8-whl., 1956	40	105	___	___	___
6561 Reel Car, 1953-56	20	60	___	___	___
6562 N Y C Gondola, 1956-58, 1966-69	7	25	___	___	___
6562-1 N Y C Gondola	3	9	___	___	___
6572 Railway Express Refrig. Car, 58-59, 63	25	60	___	___	___
6630 I R B M Rocket Launcher, 1960-64	20	75	___	___	___
6636 Alaska Quad Hopper, 1959-60	17	40	___	___	___
6640 U S M C Rocket Launcher, 1960	75	225	___	___	___
6646 Lionel Lines Stock Car, 1957	10	30	___	___	___
6650 I R B M Rocket Launcher, 1959-63	20	45	___	___	___
6650-80 Missile Car, 1960	1	4	___	___	___
6651 U S M C Cannon Car, c. 1960-61	40	100	___	___	___
6656 Lionel Lines Stock Car, 1949-55	5	15	___	___	___
6657 Rio Grande SP Type Caboose, 1957-58	50	125	___	___	___
6660 Flatcar with crane, 1958	20	65	___	___	___
6670 Flatcar with boom, 1959-60	20	50	___	___	___
6672 Santa Fe Refrigerator Car, 1954-56	20	50	___	___	___
6736 Detroit & Mack. Quad Hop., 1960-62	17	35	___	___	___
6763-85 Lehigh Valley Hopper	5	9	___	___	___
6800 Flatcar with airplane, 1957-60	75	150	___	___	___
6801 Flatcar with boat, 1957-60	20	50	___	___	___
6802 Flatcar with bridge, 1958-59	10	25	___	___	___
6803 Flatcar with tank and truck, 1958-59	50	120	___	___	___
6804 Flatcar with U S M C trucks, 1958-59	50	120	___	___	___
6805 Atomic Disposal Flatcar, 1958-59	35	100	___	___	___
6806 Flatcar with U S M C trucks, 1958-59	25	85	___	___	___
6807 Lionel Flatcar with boat, 1958-59	35	75	___	___	___
6808 Flatcar with U S M C trucks, 1958-59	60	125	___	___	___
6809 Flatcar with U S M C trucks, 1958-59	60	125	___	___	___

Postwar	Good	Exc	Color	Cond	$
6810 Flatcar with trailer, 1958	15	35			
6812 Track Maintenance Car, 1959	15	100			
6814 Lionel Work Caboose, 1959-61	30	125			
6816-100 Allis Chalmers Tractor, 1956-60	25	75			
6816 Flatcar with bulldozer, red car, 1959-60	75	275			
6816 Flatcar w/bulldozer, black car, 1959-60	200	600			
6817 Flatcar with scraper, black, 1959-60	200	600			
6817 Flatcar with scraper, red, 1959-60	60	275			
6818 Transformer Car, 1958	10	40			
6819 Flatcar with helicopter, 1959-60	25	75			
6820 Flatcar with helicopter, 1960-61					
(A) Light blue-painted flatcar	20	60			
(B) Darker blue flatcar	40	150			
6821 Flatcar with Crates, 1959-60	10	20			
6822 Searchlight Car, 1961-69	20	50			
6823 Flatcar with I R B M missiles, 1959-60	20	65			
6824 U S M C Work Caboose, 1960	50	160			
(6824) Rescue Unit Work Caboose, c. 1963	50	125			
6825 Flatcar with bridge, 1959-62	15	45			
6826 Flatcar with trees, 1959-60	30	110			
6827 Flatcar with steam shovel, 1960-63	40	90			
6827-100 Harnischfeger Shovel, 1960	5	9			
6828 Flatcar with crane, 1960-63, 1966	40	90			
6828-100 Harnischfeger Crane, 1960	15	45			
6830 Flatcar with submarine, 1960-61	40	90			
6844 Flatcar with missiles, 1959-60					
(A) Black plastic flatcar	20	60			
(B) Red unpainted plastic flatcar	—	800			
6427 Virginian (See 6427-60)					
63132 (See 3464)					
64173 (See 6427 Lionel)					
65400 (See 2454 or 6454)					
81000 (See 6417 PRR)					
96743 (See 6454)					
159000 (See 3464)					
477618 (See 2457 or 2472)					
536417 (See 6417 PRR)					
546446 (See 6446-1 or 6446-25)					
576427 (See 6427-500)					
641751 (See 6417-50)					
576419 (See 6419-57)					
336155 (See 3361-55)					
576419 (See 6419-57)					
A Transformer, 90 watts, 1947-48	8	10			
CTC Lockon (O and O27), 1947-69	.20	.75			
ECU-1 Electronic Control Unit, 1946	10	40			
KW Transformer, 190 watts	75	95			
LTC Lockon (O and O27), 1950-69	3	5			
LW Transformer, 125 watts, 1955-56	30	40			
OC Curved Track (O), 1945-61	.15	.40			
OCS Curved Insulated Track (O), 1946-50	1	2			
OC1/2 Half Sec. Curve Track (O), 1945-66	.20	.40			
OC18 Steel Pins (O), 1945-59	.02	.05			
OC51 Steel Pins /dozen (O), 1961	.20	.50			

Postwar	Good	Exc	Color	Cond	$
OS Straight Track (O), 1945-61	.20	.40	___	___	___
OSS Straight Insulated Track, 1946-50	1	2	___	___	___
OTC Lockon Track (O and O27)	1	3	___	___	___
Q Transformer, 75 watts, 1946	8	12	___	___	___
R Transformer, 110 watts, 1946-47	15	20	___	___	___
RW Transformer, 110 watts, 1948-54	30	40	___	___	___
RCS Remote Control Track (O), 1945-48	1	3	___	___	___
SP Smoke Pellets, bottle, 1948-69	—	10	___	___	___
SW Transformer, 130 watts, 1961-66	40	50	___	___	___
T011-43 Fiber Pins /dozen (O), 1962-66	.20	.40	___	___	___
TW Transformer, 175 watts, 1953-60	40	50	___	___	___
TOC Curved Track (O), 1962-66, 68-69	.15	.40	___	___	___
TOC1/2 Half Sec. Str. Track (O), 1962-66	.20	.40	___	___	___
TOC51 Steel Pins /dozen (O), 1962-69	.20	.50	___	___	___
TOS Straight Track (O), 1962-69	.20	.40	___	___	___
UCS Remote Control Track (O), 1945-69	3	8	___	___	___
UTC Lockon (O, O27, Standard), 1945	.25	.75	___	___	___
V Transformer, 150 watts, 1946-47	40	50	___	___	___
VW Transformer, 150 watts, 1948-49	80	100	___	___	___
Z Transformer, 250 watts, 1945-47	60	75	___	___	___
ZW Transformer, 250 watts, 1948-49	125	185	___	___	___
ZW Transformer, 275 watts, 1950-60	125	185	___	___	___
No Number Caboose, SP Type, olive drab	1	2	___	___	___
No Number Caboose (See 6057)					
No Number Caboose, red, SP4 (See 6167-50)					
No Number Caboose, yellow, SP4 (See 6167-50)					
No Number Caboose, work, yellow (See 6120)					
No Number Flatcar (See 1877)					
No Number Flatcar, olive drab (See 6401-50)					
No Number Gondola, blue	1	2	___	___	___
No Number Gondola, green	1	2	___	___	___
No Number Hopper, gray (See 6076)					
No Number Hopper, yellow (See 6176)					
No Number Rolling Stock (See 1877, 3309, 3349)					
No Number Turbo Missile Car (See 3309)					
No Number Turbo Missile Car (See 3349)					

SECTION III
LIONEL 1970-1989

		Exc	LN	Color	Cond	$
3	Union Pacific 4-4-0, 1981 (See 8104)					
3	W & A R R 4-4-0 (See 8701)					
0104	Wheat Thins Reefer, 1984	NRS	NRS	___	___	___
0124	Oreo Reefer, 1984	NRS	NRS	___	___	___
303	Stauffer Chemical Tank, 1985	50	75	___	___	___
350E	Hiawatha Locomotive (See O Gauge Classics)					
484	(See 8587)					
491	N & W Diner (See 7203)					
0511	T C A St. Louis Baggage Car, 1981	—	75	___	___	___
0512	N Y C Toy Fair Reefer, 1981	125	175	___	___	___
530	(See 8378)					
550C	Curved Track, 1970	.75	1.25	___	___	___
550S	Straight Track, 1970	.75	1.25	___	___	___
577	Norfolk & Western (See 9562)					
578	Norfolk & Western (See 9563)					
579	Norfolk & Western (See 9564)					
580	Norfolk & Western (See 9565)					
581	Norfolk & Western (See 9566)					
582	N & W Vista Dome (See 9567)					
611	Norfolk & Western (See 8100)					
634	Santa Fe NW-2 Switcher, c. 1970	55	100	___	___	___
659	Chicago & Alton (See 8101)					
665E	Johnny Cash Blue Train Steam, 1971	NRS	NRS	___	___	___
672	(See 8615)					
779	(See 8215)					
0780	Lionel R R Club Boxcar, 1982	75	100	___	___	___
0781	Lionel R R Club Flatcar w/2 vans, 1983	75	100	___	___	___
0782	Lionel R R Club Tank Car, 1985	75	100	___	___	___
783	New York Central (See 8406)					
0784	Lionel R R Club Quad Hopper, 1984	50	60	___	___	___
784	(See 8606)					
882	Hiawatha Front Coach (See O Gauge Classics)					
883	Hiawatha Center Coach (See O Gauge Classics)					
884	Hiawatha Observation (See O Gauge Classics)					
1018	1979 T C A Ceremony Boxcar, 1979	80	95	___	___	___
1100	Happy Huff 'n Puff, 1975	30	40	___	___	___
1200	Gravel Gus, 1975	35	55	___	___	___
1203	B & M NW-2 Switcher, 1972	—	80	___	___	___
1278	P H D L O T S 1989 Boxcar	—	30	___	___	___
1300	Gravel Gus Junior, 1975	NRS	NRS	___	___	___
1355	Train Display Case, Set 1355, 1983	—	40	___	___	___
1359	Train Display Case, Set 1355, 1983	—	40	___	___	___
1400	Happy Huff 'n Puff Junior, 1975	40	65	___	___	___
1776	B & A GP-9 powered (w/cab.), 76	100	125	___	___	___
1776	N & W N5C Caboose, 1976	20	30	___	___	___
1776	N & W GP-9 powered, 1976	100	125	___	___	___
1776	Seaboard U36B powered, 1976	85	125	___	___	___
1970	(See 8615)					
1973	T C A Convention Auto Carrier, 1973	45	55	___	___	___
1973	T C A Bicentennial Passenger Car	40	60	___	___	___

		Exc	LN	Color	Cond	$
1974	T C A Bicentennial Passenger Car	40	60			
1975	T C A Bicentennial Passenger Car	40	60			
1976	Seaboard U36B and 3 Cars above	—	350			
1980	The Rock GP-20, L C C A, 1980 (See 8068)					
1983	Nabisco Brands NW-2 Switcher, 1983	—	200			
1984	Ritz Boxcar, 1984	NRS	NRS			
1987	Mopar Steam 4-4-2	—	50			
1987	Mopar Express Gondola, 1987	—	50			
1987	Mopar Boxcar	—	50			
1987	Mopar Flatcar with Vans	75	100			
1987	Mopar Caboose	—	50			
2110	Graduated Trestle, 1971-83, 1985-88	10	12			
2111	Elevated Trestle, 1971-83, 1985-88	10	12			
2113	Tunnel Portals (2), 1984-87	5	6			
2115	Dwarf Signal, 1984-87	11	13			
2117	Block Target Signal, 1984-88	17	21			
2122	Extension Bridge, 1977-83, 1986-87	30	45			
2125	Whistling Freight Station, 1971	40	60			
2126	Whistling Freight Shed, 1976-83, 85-87	20	25			
2127	Diesel Horn Shed, 1976-83, 1985-87	20	25			
2128	Automatic Switchman, 1983, 1985-86	25	30			
2129	Illuminated Freight Sta., 1983, 1985-86	20	30			
2133	Illuminated Freight Station, 1972-83	20	25			
2140	Automatic Banjo Signal, 1970-84	15	20			
2145	Automatic Gateman, 1970-84	20	30			
2146	Crossing Gate, 1970-71	15	20			
2151	Automatic Semaphore, 1978-84	20	25			
2152	Crossing Gate, 1972-86	20	25			
2154	Highway Flasher, 1970-83, 1985-87	15	20			
2156	Station Platform, 1970-71	35	60			
2162	Gate & Signal, 1970-83, 1985-87	20	25			
2163	Block Target Signal, 1970-78	20	25			
2170	Street Lamps (3), 1970-83, 1985-87	20	25			
2171	Goose Neck Lamps (2), 1980-84	20	25			
2175	Sandy Andy Kit, 1976-79	20	40			
2180	Road Sign Set, 1977-83, 1985-88	2	3			
2181	Telephone Poles, 1977-81, 83, 85-88	2	4			
2195	Floodlight Tower, 1970-72	40	50			
2199	Microwave Tower, 1972-75	30	50			
2214	Girder Bridge, 1970-83, 1985-87	5	8			
2256	Station Platform, 1973-81	6	10			
2256	Station Platform T C A, 1975	20	30			
2260	Bumpers (Set of 3), 1970-73	25	40			
2260	Bumpers (Set of 3), 1986	—	3			
2280	Bumpers (Set of 3), 1973-80, 1983	3	5			
2281	Black Bumpers (Set of 3), 1983	15	18			
2282	Bumpers (Pair), 1983	20	25			
2283	Die-cast Bumpers, red (pair), 1984-88	10	12			
2290	Lighted Bumpers (pair), 1974-83, 85-86	6	8			
2292	Station Platform, 1985-87	4	6			
2300	Operating Oil Drum Loader, 1985-87	75	95			
2301	Operating Sawmill, 1981-1985	55	75			
2302	U P Gantry Crane Kit, 1981-82	15	25			
2303	Santa Fe Gantry Crane Kit, 1981-82	20	40			
2305	Operating Oil Derrick, 1981-86	75	120			

		Exc	LN	Color	Cond	$
2306	Oper. Icing Station and Car, 1982-83	145	200			
2307	Billboard Light, 1982-83, 1985-86	15	20			
2308	Animated Newsstand, 1982-83	90	120			
2309	Mechanical Gate, 1982-83, 1985-88	2	4			
2310	Gate and Signal, 1973-75	3	5			
2311	Mechanical Semaphore, 1982-84, 86-88	2	4			
2312	Mechanical Semaphore, 1973-75	5	6			
2313	Floodlight Tower, 1975-84, 1986	20	25			
2314	Searchlight Tower, 1975-85	20	30			
2315	Operating Coaling Station, 1984-86	95	135			
2316	Oper. N & W Gantry Crane, 1984	90	120			
2317	Operating Drawbridge, 1975-81	40	70			
2318	Operating Control Tower, 1984-86	55	65			
2319	Watchtower, illuminated, 1975-80	20	30			
2320	Flagpole Kit, 1983-88	3	5			
2321	Operating Sawmill, 1986-87	60	100			
2323	Operating Freight Station, 1985-87	50	65			
2324	Operating Switch Tower, 1985-87	55	65			
2383	Illuminated Bumpers, die-cast, 1987	—	12			
2390	Lionel Mirror, 1982	50	65			
2494	Rotary Beacon, 1972-74	40	50			
2671	Cleveland T C A, 1975	75	125			
2709	Rico Station Kit, 1981-88	15	25			
2710	Billboards (5), 1970-84	5	6			
2714	Tunnel, 1975-77	5	7			
2716	Short Extension Bridge, 1988	—	6			
2717	Short Extension Bridge, 1977-87	3	5			
2718	Barrel Platform Kit, 1977-83, 1985	2	4			
2719	Signal Tower Kit, 1977-83, 1986-87	2	4			
2719	Watchman's Shanty Kit, 1985-87	2	4			
2720	Lumber Shed Kit, 1977-83, 1986-87	2	4			
2721	Operating Log Mill, 1979	NRS	NRS			
2722	Barrel Loader, 1979	3	5			
2723	Barrel Loader, 1984	3	5			
2729	Water Tower Building Kit, 1985			Not Manufactured		
2783	Manual Freight Station Kit, 1981-83	7	10			
2784	Freight Platform Kit, 1981-83, 1985-88	6	9			
2785	Engine House Kit, 1974-77	30	40			
2786	Freight Platform Kit, 1974-77	4	7			
2787	Freight Station Kit, 1974-77, 1983	7	10			
2788	Coaling Station Kit, 1975-77	25	40			
2789	Water Tower Kit, 1975-80, 1985	7	12			
2790	Building Kit Assortment, 1983	NRS	NRS			
2791	Cross Country Set, 1970-71	10	25			
2792	Layout Starter Pak, 1980-83	10	25			
2792	Whistle Stop Set, 1970-71	15	25			
2793	Alamo Junction Set, 1970-71	—	25			
2796	Grain Elevator Kit, 1977	60	75			
2797	Rico Station Kit, 1976	30	45			
2900	Lockon, 1970-87	1	1.25			
2901	Track Clips (12), 1970-87	2	5			
2905	Lockon and Wire, 1972-87	.85	1.50			
2909	Smoke Fluid, 1977-87	1.50	3			
2927	Maintenance Kit, 1970-71, 1977-87	3	5			
2951	Track Book, 1976-80, 1983, 1985-86	.75	2			

		Exc	LN	Color	Cond	$
2952	Track Accessories Manual, 1985	.75	1	___	___	___
2953	Track & Acc. Man. 1977-83, 1985-86	1.25	2	___	___	___
2960	Lionel 75th Anniversary Book, 1976	8	10	___	___	___
2980	Magnetic Conversion Coupler, 1979	1	2	___	___	___
2985	The Lionel Train Book, 1986	—	7	___	___	___
3080	Train Display Layout	175	250	___	___	___
3100	Great Northern 4-8-4 Steam, 1981	450	525	___	___	___
3764	Kahn Boxcar, L O T S, 1981	25	50	___	___	___
4044	Transformer 45-W, 1970-71	3	5	___	___	___
4045	Safety Transformer, 1970-71	3	5	___	___	___
4050	Safety Transformer, 1972-79	3	5	___	___	___
4060	Power Master, Trans., 1980-83, 85-88	12	30	___	___	___
4065	Commando Set, DC Trans., 1986	15	25	___	___	___
4065	DC Hobby Transformer, 1981-83	3	4	___	___	___
4090	Power Master, Trans., 1970-81, 1983	40	55	___	___	___
4125	Transformer 25-W, 1972	3	4	___	___	___
4150	Trainmaster Transformer, 1976-78	6	12	___	___	___
4250	Trainmaster Transformer	6	12	___	___	___
4449	(See 8307)					
4501	(See 8309)					
4651	Trainmaster Transformer, 1978-79	2	3	___	___	___
4690	Trans., Type MW, Solid-state, 1987-88	—	90	___	___	___
4851	Transformer, 1987-88	3	4	___	___	___
4870	DC Hobby Transformer & Throttle Controller, 1977-78	3	4	___	___	___
4935	(See 8150)					
5001	Curved Track (O27), 1985	NA	NA	___	___	___
5012	Curved Track (O27), 1980-88	2.20	3.50	___	___	___
5013	Curved Track (O27)	.50	.55	___	___	___
5014	Half-Curved (O27), 1980-88	.50	.75	___	___	___
5016	3-foot Straight Track (O27), 1987	—	2.25	___	___	___
5017	Straight Track (O27), 1980-88	—	3.50	___	___	___
5018	Straight Track (O27)	.50	.75	___	___	___
5019	Half-Straight Track (O27), 1986-88	.50	.75	___	___	___
5020	90° Crossover (O27), 1970-88	3.50	4.75	___	___	___
5021	Manual Left Switch (O27), 1986-88	8	13	___	___	___
5022	Manual Right Switch (O27), 1986-88	8	13	___	___	___
5023	45° Crossover (O27), 1970-88	4	6.50	___	___	___
5024	Straight Track	—	2.25	___	___	___
5025	Manumatic Uncoupler, 1971-75	1	2	___	___	___
5027	Manual Switches, pair (O27)	15	25	___	___	___
5030	Track Expander Set, 1972	—	25	___	___	___
5030	Layout Set, 1978-80, 1983, 1985	17	25	___	___	___
5033	Curved Track (O27), 1986-88	.30	.75	___	___	___
5038	Straight Track (O27), 1986-88	.30	.75	___	___	___
5041	Insulator Pins (dz.) (O27), 1986-88	.50	1	___	___	___
5042	Steel Pins (8/pack) (O27), 1986-88	.40	.75	___	___	___
5044	Curved Track Ballast (O42), 1987-88	—	2.25	___	___	___
5045	Wide Radius Track Ballast (O27), 87-88	—	2.25	___	___	___
5046	Curved Track Ballast (O27), 1987-88	—	2.25	___	___	___
5047	Straight Track Ballast (O27), 1987-88	—	2.25	___	___	___
5049	Curved Track (O42)	—	1.10	___	___	___
5090	Manual Switches (3 pair), 1983, 1985	60	80	___	___	___
5100	(See 18001)					
5113	Wide Rad. Curved Track (O27), 86-87	1.10	1.50	___	___	___

		Exc	LN	Color	Cond	$
5121	Left R.C. Switch (O27), 1986-88	15	20			
5122	Right R.C. Switch (O27), 1986-88	15	20			
5125	R.C. Switches, pair (O27), 1986	25	40			
5132	R.C. Right Switch (O27), 1986-88	33	40			
5133	R.C. Left Switch (O), 1986-88	33	40			
5149	R.C. Uncoupling Track (O), 1986-88	3	7			
5165	O72 Wide Rad. Rem. Switches, 87-88	—	50			
5166	O72 Wide Rad. Rem. Switches, 87-88	—	50			
5167	Remote Switch (O42), Right	—	25			
5168	Remote Switch (O42), Left	—	25			
5193	R.C. Switches (3 pair) (O27), 1983	90	110			
5484	T C A Steam 4-64, 1985	400	450			
5500	Straight Track (O), 1971, 1983-88	.75	1.25			
5501	Curved Track (O), 1971, 1983-88	.75	1.25			
5502	Remote Control Track, 10", 1971	5	8			
5504	Half-Curved (O), 1983, 1985-88	—	1.05			
5505	Half-Straight (O), 1983, 1985-88	—	1.05			
5510	Curved Track (O)	.75	1.25			
5520	90° Crossover (O), 1971	5	7			
5522	3-foot Straight Track, 1987	—	3.75			
5523	Straight Track, 40-inch	—	5			
5530	Remote Uncoupling Section (O), 86-88	11	13			
5540	90° Crossover (O), 1981-88	7	8			
5543	Insulator Pins (12) (O), 1970, 1986-88	.75	1			
5545	45° Crossover (O), 1982-88	—	13			
5551	Steel Pins (12/pack) (O), 11970, 986-88	.75	1			
5560	Wide Rad. Curved Track, Ballast, 87-88	—	2.25			
5561	Wide Rad. Curved Track, Ballast, 87-88	—	2.25			
5562	Wide Rad. Curved Track, Ballast, 87-88	—	2.25			
5572	Wide Rad. Curved Track (O72), 86-88	1.25	2			
5600	Curved Track, 1973	—	1.75			
5601	Card of Four Curved Track, 1973	—	7			
5602	Card of Four Roadbed Ballast, Curved Track, 1973-74	4	7			
5605	Straight Track, 1973	1	2			
5606	Card of Four Straight Track, 1973	4	8			
5607	Card of Four Roadbed Ballast, Straight Track, 1973	4	8			
5620	Left Manual Switch, 1973-74	—	25			
5625	Left Remote Switch, 1973	NRS	NRS			
5630	Right Manual Switch, 1973-74	—	25			
5635	Right Remote Switch, 1973	NRS	NRS			
5640	Left Switch Roadbed Pieces (2), 1973	—	10			
5650	Right Switch Roadbed Pieces (2), 1973	—	8			
5655	Lockon, 1973	—	1.50			
5660	Terminal Track w/Lockon, 1973	—	3.50			
5700	Oppenheimer Woodside Reefer, 1981	25	45			
5701	Dairymen's League Reefer, 1981	15	20			
5702	Nat. Dairy Despatch Reefer, 1981	15	20			
5703	No. American Despatch Reefer, 1981	15	20			
5704	Budweiser Woodside Reefer, 1981	20	30			
5705	Ball Glass Jars Woodside Reefer, 1981	15	20			
5706	Lindsay Bros. Woodside Reefer, 1981	20	25			
5707	American Refrigeration Reefer, 1981	15	20			
5708	Armour Woodside Reefer, 1982-83	12	15			

		Exc	LN	Color	Cond	$
5709	R E A Reefer, Woodside, 1982- 83	30	45			
5710	Canadian Pacific Reefer, Wood., 82-83	12	15			
5711	Commercial Exp. Reefer, Wood., 82-83	12	15			
5712	Lionel Woodside Reefer, 1982	200	300			
5713	Cotton Belt Reefer, Woodside, 1983	15	20			
5714	Michigan Central Reefer, 1983	15	20			
5715	Santa Fe Reefer, orange, 1983	15	20			
5716	Vermont Central Reefer, 1983	15	20			
5717	A T & S F Bunk Car, gray, 1984	40	45			
5718	(See 9849)					
5719	Canadian National Reefer, 1984	15	20			
5720	Great Northern Reefer, 1985	85	125			
5721	SOO Line Woodside Reefer, 1985	15	20			
5722	Nickel Plate Reefer Woodside, 1985	15	20			
5724	Pennsylvania Bunk Car, 1985	20	25			
5726	Southern Bunk Car, 1985	30	40			
5727	U. S. Marines Bunk Car, 1985	15	20			
5728	Canadian Pacific Bunk Car, 1986	15	20			
5730	Strasburg R R Reefer, 1985-86	15	20			
5731	L & N Woodside Reefer, 1985-86	12	15			
5732	Central R R of N J Reefer, 1985-86	12	15			
5733	Lionel Lines Bunk Car, 1986	40	50			
5734-85	R E A T C A Reefer, 1985	150	225			
5735	N Y C Bunk Car, gray, 1985-86	35	45			
5739	B & O Tool Car, Gray, 1986	40	50			
5745	A T & S F Bunk Car, 1986	40	50			
5760	A T & S F Tool Car, 1986	40	50			
5823	45° O Gauge Crossover	4	6			
5900	AC/DC Converter, 1979-81, 1983	4	6			
5745	A T S F Bunk Car, 1986	40	50			
5760	A T S F Tool Car, 1986	40	50			
5900	AC/DC Converter, 1979-81, 1983	4	6			
6014-900	Frisco Boxcar, 1975	50	75			
6076	Lehigh Valley Short Hopper, 1970	NRS	NRS			
6100	Ontario Northland Quad, 1981-82	25	30			
6101	B N Quad Hopper, green, 1981, 83	15	25			
6102	G N Quad (FARR #3), 1981	35	50			
6103	C N Quad (FARR #4), 1981	25	35			
6104	Southern Quad w/Coal, 1984-85	45	55			
6105	Reading Oper. Short Hopper, 1982	30	40			
6106	N & W Quad Hopper, gray, 1982	35	45			
6107	Shell Quad Hopper, yellow, 1982	15	20			
6109	C & O Operating Short Hopper, 1983	25	40			
6110	MoPac Quad Hopper, black, 1983	15	25			
6111	L & N Quad Hopper, gray, 1983	15	25			
6112	Commonwealth Edison Hopper, 1983	80	120			
6113	Illinois Central Short Hopper, 1985	10	15			
6114	C & N W Quad Hopper, 1983	45	65			
6115	Southern Quad with Coal, 1983	—	20			
6116	SOO Line Ore Car, 1985	25	35			
6117	Erie Operating Short Hopper, 1985	25	35			
6118	Erie-Lackawanna Quad, 1985	35	40			
6122	Penn Central Ore Car, 1985	25	35			
6123	Penn Quad Hopper, gray, 1985	35	45			
6124	D & H Quad Hopper, red, 1984	15	25			

		Exc	LN	Color	Cond	$
6126	Canadian National Ore Car, 1986	17	25			
6127	(See 5735)					
6127	Northern Pacific Ore Car, 1986	17	25			
6131	Illinois Terminal Quad Hopper, 1985-86	17	25			
6134	B N A C F Short Hopper, 1986	150	200			
6135	C & N W A C F Hopper, 1986	150	200			
6137	N K P Road Hopper, 1986	10	15			
6138	B & O Quad Hopper, 1986	20	30			
6150	N & W Short Hopper, black, 1970	125	150			
6150	Santa Fe Hopper, blue, 1985-86	10	15			
6177	Reading Short Hopper, brown, 1986	10	15			
6200	F E C Gondola, red, 1981	15	30			
6201	Union Pacific Gondola, 1982-83	15	20			
6202	Western MD Gondola w/ coal, 1982	25	35			
6205	Canadian Pacific Gondola, 1983	25	30			
6206	C & I M Gondola, red, 1985	10	15			
6207	Southern Gondola, black (O27), 1984	6	8			
6208	B & O "Chessie" Gondola, 1983-84	25	30			
6209	N Y C Gondola w/coal (Std. O), 1985	50	75			
6210	Erie-Lackawanna Gondola, 1985	20	30			
6211	C & O Gondola (O27), 1984	—	10			
6214	Lionel Lines Gondola, 1985	25	40			
6230	Erie Reefer (Std. O), 1986	135	155			
6231	Railgon Gondola (Std. O), 1986	135	155			
6232	Illinois Central Boxcar (Std. O), 1986	135	155			
6233	C P Flatcar with stakes (Std. O), 1986	135	155			
6234	B N Boxcar, green (Std. O), 1985	—	40			
6235	B N Boxcar, green (Std. O), 1985	—	40			
6236	B N Boxcar, green (Std. O), 1985	—	40			
6237	B N Boxcar, green (Std. O), 1985	—	40			
6238	B N Boxcar, green (Std. O), 1985	—	40			
6239	B N Boxcar (Std. O), 1986	—	40			
6251	N Y C Oper. Coal Dump Car, 1985-86	15	20			
6254	N K P Road Gondola, 1986	10	12			
6258	A T S F Gondola, 1985-86	—	6			
X6260	N Y C Gondola, 1985	15	18			
6272	A T & S F Long Gondola, 1986	20	25			
6300	Corn Products 3-D Tank Car, 1981	20	35			
6301	Gulf 1-D Tank Car, 1981-82	20	25			
6302	Quaker State 3-D Tank Car, 1981-82	30	40			
6304	G N 1-D Tank Car (FARR #3), 1981	45	65			
6305	British Columbia 1-D Tank Car, 1981	45	65			
6306	Southern 1-D Tank Car, 1984-85	40	55			
6307	Pennsylvania 1-D Tank Car, 1985	40	55			
6308	Alaska 1-D Short Tank Car, 1982	10	15			
6310	Shell 2-D Tank Car (O27), 1983	12	15			
6312	C & O 2-D Tank Car (O27), 1984	15	20			
6313	Lionel Lines 1-D Tank Car, 1985	40	65			
6314	B & O 3-D Tank Car, black, 1986	40	65			
6315	T C A 18th Convention Tank, 1972	75	95			
6317	Gulf 2-D Tank Car (O27), 1984	15	20			
6323	Va. Chem. 1-D Tank, L C C A, 1985	40	55			
6325	(See 6579)					
6326	N & B Bay Window Caboose	—	30			
6357	Frisco 1-D Tank Car, black, 1983	40	60			

		Exc	LN	Color	Cond	$
6401	Virginian B/W Caboose, 1981	20	25	___	___	___
6403	Amtrak Vista Dome, 1976-77	30	50	___	___	___
6404	Amtrak Pullman, 1976-77	30	50	___	___	___
6405	Amtrak Pullman, 1976-77	30	50	___	___	___
6406	Amtrak Observation, 1976-77	30	50	___	___	___
6410	Amtrak Pullman, 1977-78	30	50	___	___	___
6411	Amtrak Pullman, 1977-78	30	50	___	___	___
6412	Amtrak Vista Dome, 1977-78	30	50	___	___	___
6420	Reading Maintenance Caboose, 1981-82	15	20	___	___	___
6421	Cowen B/W Caboose, 1982	25	30	___	___	___
6422	Duluth Missabe B/W Caboose, 1981-82	20	25	___	___	___
6425	Erie-Lack. Bay Window Caboose, 1984	20	25	___	___	___
6426	Reading Maintenance Caboose, 1982-83	10	12	___	___	___
6427	Burlington North. Maint. Cab., 1983	8	10	___	___	___
6428	C & N W Maintenance Cab., 1985	12	15	___	___	___
6430	Santa Fe SP Caboose, 1983	6	8	___	___	___
6431	Southern B/W Cab. (FARR #4) , 83-84	35	45	___	___	___
6432	Union Pacific SP Caboose, 1981	10	12	___	___	___
6433	C P Bay Window Caboose, 1981-82	40	60	___	___	___
6434	S S R R, 1983	6	8	___	___	___
6435	Commando Set Security Cab., 1983	—	20	___	___	___
6438	G N B/W Cab. (FARR #3), 1981	35	50	___	___	___
6439	Reading B/W Caboose, 1985-86	20	30	___	___	___
6441	Alaska B/W Caboose, 1982-83	25	30	___	___	___
6446-25	N & W Quad Hopper, 1970	150	175	___	___	___
6446-50	Minneapolis & St. Louis Box., 1970	—	500	___	___	___
6449	Wendy's N5C Caboose, 1981-82	40	50	___	___	___
6464-500	Timken Boxcar, 1970	100	150	___	___	___
6464-1970	T C A Special Boxcar, 1970	125	160	___	___	___
6464-1971	T C A Special Boxcar, 1971	200	250	___	___	___
6481	Rio Grande SP Caboose, 1983	6	8	___	___	___
6482	Nibco Express SP Caboose, 1982	—	75	___	___	___
6483	L C C A Caboose, 1982	—	50	___	___	___
6485	Chessie Caboose, 1984-85	—	10	___	___	___
6486	Southern SP-type Caboose, 1985	8	12	___	___	___
6491	Erie-Lack. Transfer Cab., 1985-86	—	12	___	___	___
6493	L & C B/W Caboose, 1986-87	20	25	___	___	___
6494	A T & S F 4-Whl. Bobber Cab., 1986	8	10	___	___	___
6496	A T & S F Work Caboose, 1986	—	30	___	___	___
6500	Lionel Depressed Flatcar, 1982 (See 9233)					
6504	L A S E R Helicopter, 1981-82	12	30	___	___	___
6505	L A S E R Guidance, 1981-82	12	30	___	___	___
6506	L A S E R Security Car, 1981-82	15	30	___	___	___
6507	L A S E R Missile Car, 1981-82	12	30	___	___	___
6508	Canadian Pacific Crane, 12-whl., 1981	45	70	___	___	___
6509	Lionel 16-whl. Flatcar w/girders, 1981	60	85	___	___	___
6510	Union Pacific Crane, 1982	50	75	___	___	___
6515	U P Flatcar, 1986	—	10	___	___	___
6521	N Y C Flatcar w/stakes (Std. O), 1985	35	50	___	___	___
6522	C & N W Searchlight Car, 1984-85	20	25	___	___	___
6524	Erie-Lackawanna Crane, 12-whl., 1985	45	70	___	___	___
6526	U. S. Marines Searchlight Car, 1985	15	20	___	___	___
6529	N Y C Searchlight Car, gray, 1985-86	20	25	___	___	___
6531	Express Mail Flatcar w/vans, 1985-86	15	25	___	___	___
6560	Bucyrus Erie Crane Car, 8-whl., 1971	125	200	___	___	___

		Exc	LN	Color	Cond	$
6561	Flatcar with Cruise Missile, 1983-84	—	25			
6562	Flatcar with Barrels, 1983-84	—	25			
6564	Flatcar with 2 U S M C Tanks, 1983-84	—	25			
6567	Ill. Central Gulf Crane, L C C A, 1986	60	75			
6573	Redwood Valley Express Flatcar with log dump, 1984-85	—	15			
6574	Redwood Valley Exp. Short Crane, 1984	—	15			
6575	Redwood Valley Express Flat., 1984-85	—	15			
6576	A T S F Short Crane, 1985-86	—	12			
6579	N Y C Crane, black, 8-wheel, 1985-86	30	50			
6582	T T O S Flatcar, 1986	NRS	NRS			
6585	Penn. Flatcar with fences, yellow, 1986	—	10			
6587	W & A General, 1986	20	30			
6593	A T & S F Crane, 1986	—	50			
6670	(See 9378)					
6700	U P/S P Reefer, or., 1982-83 (see 2306)	50	65			
6900	N & W Extended Vision Cab., 1982	100	145			
6901	Ont. Northland Ext. Vision Cab., 1982	50	75			
6903	Santa Fe Extended Vision Cab., 1983	100	125			
6904	U P Extended Vision Caboose, 1983	100	150			
6905	Nickel Plate Ext. Vis. Caboose, 1984	75	100			
6906	Erie-Lackawanna Ext. Vis. Cab., 1985	75	100			
6907	N Y C Woodside Caboose, 1986	150	200			
6908	Pennsylvania N5C Caboose, 1985	65	95			
6910	N Y C Extended Vision Cab., 1985	85	115			
6912	Redwood Valley Express Cab., 1983-84	—	20			
6913	B N Extended Vision Caboose, 1986	85	115			
6916	N Y C Work Caboose, gray, 1985-86	12	15			
6917	J C Ext. Vis. Caboose, 1986	40	50			
6918	B & O SP-type Caboose, 1986	—	15			
6919	Nickel Plate Caboose, 1986	—	10			
6920	B & A Woodside Caboose, 1986	100	200			
6921	Penn SP-type Caboose, 1986	—	10			
6926	T C A Ext. Vis. Caboose, 1986	80	100			
7100	Huff 'n Puff Booklet, 1975	—	.25			
7200	Streamline Quicksilver Coach, 1982-83	20	40			
7201	Streamline Quicksilver Coach, 1982-83	20	40			
7202	Quicksilver Observation, 1982-83	20	40			
7203	N & W "491" Diner, 1984	275	400			
• 7204	S P Daylight Diner, 1984	275	400			
7205	T C A Denver Combine, 1982	—	75			
7206	T C A Louisville Pullman, 1983	—	75			
7207	New York Central Diner, 1984	175	300			
7208	Pennsylvania Diner, 1984	125	225			
7210	Union Pacific Diner, 1985	125	225			
• 7211	S P Daylight Vista Dome, 1984	300	400			
7212	T C A Convention Pitts. Pass. Car, 84	—	75			
7215	B & O Old Time Coach, 1983	30	50			
7216	B & O Old Time Coach, 1983	30	50			
7217	B & O Old Time Baggage, 1983	30	50			
7220	Ill. Central Baggage Car, 1986	75	100			
7221	Ill. Central Comb. Car, 1986	75	100			
7222	Ill. Central Passenger Car, 1986	75	120			
7223	Ill. Central Passenger Car, 1986	75	120			
7224	Ill. Central Dining Car, 1986	75	120			

		Exc	LN	Color	Cond	$
7225	Ill. Central 12-Whl. Obs. Car, 1986	75	120			
7227	Wabash 12-Whl. Dining Car, 1986-87	75	100			
7228	Wabash 12-Whl. Baggage Car, 1986-87	75	100			
7229	Wabash 12-Whl. Comb. Car, 1986-87	75	100			
7230	Wabash 12-Whl. Coach Car, 1986-87	75	100			
7231	Wabash 12-Whl. Coach Car, 1986-87	75	100			
7232	Wabash 12-Whl. Obs. Car, 1986-87	75	100			
7241	W & A Coach Car, 1986	30	50			
7242	W & A Baggage Car, 1986	30	50			
7301	N & W Stock Car, brown, 1982	40	50			
7302	T & P Stock Car (O27), 1983-84	—	10			
7303	Erie-Lack. Stock Car, blue, 1985	30	35			
7304	Southern "9459" Stock Car, 1983	45	60			
7309	Southern Short Stock Car, 1985-86	10	15			
7312	W & A Stock Car, 1986	—	25			
7401	Chessie Stock Car (O27), red, 1984	—	10			
7403	L N A C Boxcar, L C C A, 1984	35	45			
7404	Jersey Central Boxcar, 1986	40	50			
7500	Lionel 75th Anniv. U36B pow., 1975	80	100			
7501	Lionel 75th Anniv. Boxcar, 1975	15	20			
7502	Lionel 75th Anniv. Reefer, 1975	15	20			
7503	Lionel 75th Anniv. Reefer, 1975	15	20			
7504	Lionel 75th Anniv. Quad Hop., 1975	20	25			
7505	Lionel 75th Anniv. Boxcar, 1975	15	20			
7506	Lionel 75th Anniv. Boxcar, 1975	15	20			
7507	Lionel 75th Anniv. Reefer, 1975	15	20			
7508	Lionel 75th Anniv. Cab., N5C, 1975	20	30			
7509	Kentucky Fried Chicken, 1981-82	15	20			
7510	Red Lobster Reefer, 1981-82	20	30			
7511	Pizza Hut Reefer, 1981-82	20	25			
7512	Arthur Treacher's Reefer, 1982	15	20			
7513	Bonanza Reefer, 1982	15	20			
7514	Taco Bell Reefer, 1982	15	20			
7515	Denver Mint Car, 1982	85	95			
7517	Philadelphia Mint Car, 1982	45	55			
7518	Carson City Mint Car, 1983	35	50			
7519	Toy Fair Reefer, 1982	150	175			
7520	Nibco Boxcar, 1982	—	500			
7521	Toy Fair Reefer, 1983	150	175			
7522	New Orleans Mint Car, 1985	25	35			
7523	Toy Fair Refrigerator, 1984	150	175			
7524	Toy Fair Reefer, 1985	175	200			
7525	Toy Fair Boxcar, 1986 (See 7925)	150	175			
7530	Dahlonega Mint Car, 1986	65	75			
7600	Frisco N5C "Spirit" Caboose, 1975-76	—	50			
7601	Delaware Boxcar, 1975-76	15	20			
7602	Pennsylvania Boxcar, 1975-76	20	25			
7603	New Jersey Boxcar, 1975-76	15	20			
7604	Georgia Boxcar, 1975-76	20	30			
7605	Connecticut Boxcar, 1975-76	20	30			
7606	Frisco N5C Caboose, 1974	20	30			
7606	Massachusetts Boxcar, 1975-76	20	30			
7607	Maryland Boxcar, 1975-76	20	30			
7608	South Carolina Boxcar, 1975-76	20	30			
7609	New Hampshire Boxcar, 1975-76	35	45			

		Exc	LN	Color	Cond	$
7610	Virginia Boxcar, 1976	100	195			
7611	New York Boxcar, 1976	60	75			
7612	North Carolina Boxcar, 1976	30	50			
7613	Rhode Island Boxcar, 1976	50	65			
7700	Uncle Sam Boxcar, 1976	60	75			
7701	Camel Boxcar, 1976-77	12	14			
7702	Prince Albert Boxcar, 1976-77	12	14			
7703	Beechnut Boxcar, 1976-77	12	14			
7704	Welcome Toy Fair Boxcar, 1976-77	175	200			
7705	Toy Fair Boxcar, 1976	300	400			
7706	Sir Walter Raleigh Boxcar, 1977-78	12	14			
7707	White Owl Boxcar, 1977-78	12	14			
7708	Winston Boxcar, 1977-78	12	14			
7709	Salem Boxcar, 1978	12	14			
7710	Mail Pouch Boxcar, 1978	15	20			
7711	El Producto Boxcar, 1978	15	20			
7712	A T & S F FARR #1 Box., yel., 1979	35	45			
7714	Northern Pacific Boxcar, 1980	12	15			
7784	T C A Museum, 1984	30	35			
7800	Pepsi Boxcar, 1977	20	30			
7801	A & W Boxcar, 1977	15	20			
7802	Canada Dry Boxcar, 1977	20	25			
7803	Trains' n Truckin' Boxcar, 1978	20	35			
7806	Season's Greetings Boxcar, 1976	125	150			
7807	Toy Fair Boxcar, 1977	150	200			
7808	N P Stock Car, 1977	40	55			
7809	Vernors Boxcar, 1978	12	18			
7810	Orange Crush Boxcar, 1978	12	18			
7811	Dr. Pepper Boxcar, 1978	12	18			
7812	T C A Houston Stock Car, 1977	35	45			
7813	Season's Greetings Boxcar, 1977	175	200			
7814	Season's Greetings Boxccar, 1978	175	200			
7815	Toy Fair Boxcar, 1978	150	175			
7816	Toy Fair Boxcar, 1979	150	175			
7817	Toy Fair Boxcar, 1980	150	175			
7900	Outlaw Car, 1982-83	11	14			
7901	Cop & Hobo Car, 1982-83	20	25			
7902	A T & S F Boxcar (O27), 1982-83	6	8			
7903	Rock Boxcar (O27), 1983	6	8			
7904	San Diego Zoo Car (O27), 1983	30	35			
7908	Tappan Cooking Boxcar (O), 1982	50	70			
7909	L & N Boxcar (O27), 1983	10	15			
7910	Chessie Boxcar (O27), 1984	10	15			
7912	Geoffrey Car, 1982-83	100	125			
7913	Turtleback Zoo Giraffe Car, 1986	25	30			
7914	Geoffrey Car, 1985	100	135			
7920	Sears Centennial Boxcar, 1985-86	50	60			
7925	Erie Lackawanna Boxcar, 1986	8	12			
7926	N K P Road Boxcar, 1986	8	11			
7930	True Test	—	60			
7931	Town House TV & Appliances Box., 86	—	50			
7932	Kay Bee Toys Boxcar, 1986	—	50			
7979	True Value Boxcar, 1986	—	50			
8001	Nickel Plate Steam, 2-6-4, 1980	50	60			
8002	Union Pacific Steam, 2-8-4, 1980	375	475			

		Exc	LN	Color	Cond	$
8003	Chessie Steam, 2-8-4, 1980	475	600	___	___	___
8004	Rock Island Steam, 4-4-0, 1980, 1982	150	170	___	___	___
8004	(See 18004)					
8005	A T & S F Steam, 4-4-0, 1980, 1982	30	40	___	___	___
8006	A C L Steam, 4-6-4 (J. C. Penney's Special), 1980	700	1000	___	___	___
8007	N Y N H & H Steam, 2-6-4, 1980	40	50	___	___	___
8008	Chessie 4-4-2 Steam, 1980	50	75	___	___	___
8010	A T & S F NW-2 Switcher, blue, 1970	40	75	___	___	___
8020	(A) Santa Fe A powered, 1970-76	50	95	___	___	___
8020	(B) Santa Fe Alco A dummy, 1970-71	35	65	___	___	___
8021	Santa Fe Alco B unit dummy, 1971-72, 1974-76	40	55	___	___	___
8022	Santa Fe Alco A powered, 1971	70	95	___	___	___
8023	Canadian National Alco A, 1970	35	40	___	___	___
8025	Canadian National Alco AA, 1971	125	200	___	___	___
8030	Illinois Central GP-9 pow., 1970-71	55	70	___	___	___
8031	Canadian Natl. GP-7 pow., 1970-71	70	125	___	___	___
8040	Nickel Plate Steam, 2-4-2, 1970-72	20	30	___	___	___
8040	Canadian Natl. Steam, 2-4-2, 1971	50	100	___	___	___
8041	N Y C Steam, 2-4-2, 1970-71	30	50	___	___	___
8042	Grand Trunk Steam, 2-4-2, 1970, 1972	30	40	___	___	___
8043	Nickel Plate Steam, 2-4-2, 1970	50	75	___	___	___
8050	D & H U36C powered, 1980	100	150	___	___	___
8051	D & H U36C dummy, 1980	50	75	___	___	___
8054/8055	C & S Burlington F-3 AA, 80	275	325	___	___	___
8056	C & N W Trainmaster, 1980-81	275	325	___	___	___
8057	Burlington NW-2 pow. Switcher, 80	75	150	___	___	___
8059	Penn F-3 B dummy, green, 1980	300	450	___	___	___
8060	Penn F-3 B dummy, tuscan, 1980	300	400	___	___	___
8061	Western MD U36C powered, 1980	100	125	___	___	___
8062	C & S Burlington F-3 B dummy, 80	100	150	___	___	___
8062	Great Northern Steam, 1970			Not Manufactured		
8063	Seaboard SD-9 powered, 1980	80	115	___	___	___
8064	F E C GP-9 powered, 1980	75	100	___	___	___
8065	F E C GP-9 dummy, 1980	50	60	___	___	___
8066	T P & W GP-20 powered, 1980	100	125	___	___	___
8067	T & P Alco A powered, 1980			Not Manufactured		
8068	The Rock GP-20 powered, 1980	125	150	___	___	___
8071	Virginian SD-18 powered, 1980	100	150	___	___	___
8072	Virginian SD-18 dummy, 1980	60	75	___	___	___
8100	N & W Steam, 4-8-4, J-1, 1981	900	1400	___	___	___
8101	Chicago & Alton Steam, 4-6-4, 1981	400	500	___	___	___
8102	Union Pacific Steam, 4-4-2, 1981-82	60	75	___	___	___
8104	Union Pacific Steam, 4-4-0, 1981	225	300	___	___	___
8111	D T & I NW-2 pow. Switcher, 1971-74	35	75	___	___	___
8140	Southern Steam, 2-4-0, 1971	25	35	___	___	___
8141	Pennsylvania Steam, 2-4-2, 1971	40	55	___	___	___
8142	C & O Steam, 4-4-2, 1971	55	65	___	___	___
8143	Milwaukee Steam, 4-4-2, 1971	30	40	___	___	___
8150	Penn GG-1 powered, green, 1981	525	650	___	___	___
8151	Burlington SD-28 powered, 1981	125	175	___	___	___
8152	Canadian Pacific SD-24 pow., 1981	125	200	___	___	___
8153	Reading NW-2 pow. Switcher, 1981	55	95	___	___	___
8154	Alaska NW-2 pow. Switcher, 1981-82	85	125	___	___	___

		Exc	LN	Color	Cond	$
8155	Monon U36B powered, 1981-82	60	75			
8156	Monon U36B dummy, 1981-82	50	55			
8157	Santa Fe Trainmaster, 1981	350	395			
8158	D M & I R GP-35 powered, 1981-82	75	125			
8159	D M & I R GP-35 dummy, 1981-82	40	50			
8160	Burger King GP-20 powered, 1981-82	85	100			
8161	L A S E R Gas Turbine, 1981-82	60	100			
8162	Ontario Northland SD-18 pow., 1981	100	125			
8163	Ontario Northland SD-18 dum., 1981	60	75			
8164	Penn F-3 B unit, green, horn, 1981	350	450			
8182	Nibco NW-2 pow. Switcher, 1982	125	175			
8190	Diesel Horn Kit, 1981	—	25			
8200	(See 18200)					
8200	Kickapoo Dockside Steam, 0-4-0, 1972	40	50			
8203	Pennsylvania Steam, 2-4-2, 1972	30	40			
8203	(See 18203)					
8204	C & O Steam, 4-4-2, 1972	60	70			
8206	N Y C Steam, 4-6-4, 1972-74	200	225			
8209	Pioneer Dockside Steam, 0-4-0, 1972-76					
(A)	With four-wheel Tender	40	60			
(B)	Without Tender	40	50			
8210	Cowen Steam, 4-6-4, 1982	350	400			
8212	Black Cave Steam, 0-4-0, 1982	30	50			
8213	Rio Grande Steam, 2-4-2, 1982-83	60	70			
8214	Pennsylvania Steam, 2-4-2, 1982-83	60	75			
8215	Nickel Plate Steam, 2-8-4, 1982	600	650			
8250	Santa Fe GP-9 pow., 1972-75, 1982	60	100			
8251-50	Horn/Whistle Controller, 1972-74	2	3			
8252	D & H Alco A unit powered, 1972	40	100			
8253	D & H Alco B unit dummy, 1972	20	50			
8254	Illinois Central GP-9 dummy, 1972	40	50			
8255	Santa Fe GP-9 dummy, 1972	40	50			
8258	Canadian National GP-7 dummy, 1972	40	70			
8261	S P F-3 B unit dummy, 1982	600	1000			
8260-8262	S P F-3 AA, sold as set	400	600			
8263	Santa Fe GP-7 powered, 1982	75	100			
8264	Canadian Pacific Snowplow, 1982	110	150			
8265	Santa Fe SD-40 powered, 1982	275	425			
8266	N & W SD-24 pow. w/horn, 1982	150	200			
8268	T & P Alco A unit powered, 1982-83	70	85			
8269	T & P Alco A unit dummy, 1982-83	30	50			
8272	Pennsylvania EP-5, 1982	225	250			
8300	(See 18300)					
8300	Santa Fe Steam, 2-4-0, 1976	20	25			
8302	Southern Steam, 2-4-0, 1973-76	25	30			
8303	Jersey Central Steam, 2-4-2, 1973-74	40	50			
8303	See (18303)					
8304	Baltimore & Ohio Steam, 4-4-2, 1975	75	100			
8304	C & O Steam, 4-4-2, 1974-77	55	65			
8304	Pennsylvania Steam, 4-4-2, 1974	65	75			
8304	Rock Island Steam, 4-4-2, 1973-74	55	65			
8305	Milwaukee Road Steam, 4-4-2, 1973	75	95			
8306	Pennsylvania Steam, 4-4-2, 1974	30	40			
8307	S P Steam, 4-8-4 GS-4 "4449", 1983	1800	2200			
8308	Jersey Central Steam, 2-4-2, 1973-74	40	50			

		Exc	LN	Color	Cond	$
8309	Jersey Central Steam, 2-4-2, 1974	40	50	___	___	___
8309	Southern Steam, 2-8-2 "4501" (FARR #4), 1984	525	600	___	___	___
8310	A T & S F Steam, 2-4-0, 1974-75	30	40	___	___	___
8310	Jersey Central Steam, 2-4-0, 1974-75	30	60	___	___	___
8310	Nickel Plate Steam, 2-4-0, 1974-75	30	60	___	___	___
8311	Southern Steam, 1973	30	40	___	___	___
8313	Santa Fe Steam, 0-4-0, 1983	15	20	___	___	___
8314	Southern Steam, 2-4-0, 1983	20	25	___	___	___
8315	B & O Steam, 4-4-0, 1983	85	120	___	___	___
8350	U. S. Steel Gas Turbine, 1974-75	20	30	___	___	___
8351	Santa Fe Alco A powered, 1973-74	50	75	___	___	___
8352	Santa Fe GP-20 powered, 1973-75	60	100	___	___	___
8353	Grand Trunk GP-7 powered, 1974-75	60	100	___	___	___
8354	Erie NW-2 pow. Switcher, 1973-75	75	90	___	___	___
8355	Santa Fe GP-20 dum. w/horn, 1973-75	60	80	___	___	___
8356	Grand Trunk GP-7 dummy, 1974-75	30	40	___	___	___
8357	Pennsylvania GP-9 powered, 1973-75	125	150	___	___	___
8358	Pennsylvania GP-9 dummy, 1973-75	35	50	___	___	___
8359	Chessie GP-7 powered, 1973	100	150	___	___	___
8360	Long Island GP-20 powered, 1973-74	75	95	___	___	___
8361	W P Alco A powered, 1973-74	50	75	___	___	___
8362	W P Alco B dummy, 1973-74	30	50	___	___	___
8363	B & O F-3 A, 1973-75	175	325	___	___	___
8364	B & O F-3 A dummy, 1973-75	100	150	___	___	___
8365/8366	C P F-3 AA set, 1973	650	950	___	___	___
8367	Long Island GP-20 dum. w/horn, 1973	75	100	___	___	___
8368	Alaska R R Motorized Unit, 1983	100	140	___	___	___
8369	Erie-Lack. GP-20 powered, 1983	70	100	___	___	___
8370/8371/8372	N Y C F-3 ABA, 1983	500	650	___	___	___
8374	Burlington NW-2 Switcher, 1983	65	100	___	___	___
8375	C & N W GP-7 powered, 1983	80	100	___	___	___
8376	U P SD-40, Magnetraction, 1983	325	495	___	___	___
8377	U. S. Switcher with decal sheet, 1983	60	100	___	___	___
8378	Wabash, Fairbanks-Morse 1983	850	1050	___	___	___
8379	Penn Motorized Fire Fighter, 1985	125	150	___	___	___
8380	Lionel Lines SD-28 powered, 1983	150	250	___	___	___
8402	Reading Steam, 4-4-2, 1984	60	75	___	___	___
8403	Chessie Steam, 4-4-2, 1984-85	60	75	___	___	___
8404	Pennsylvania 6-8-6, Steam, 1984-85	500	575	___	___	___
8406	N Y C Semi-scale Hudson Steam, 4-6-4, 1985	900	1100	___	___	___
8410	Redwood Steam, 4-4-0, 1984-85	40	60	___	___	___
8452	Erie Alco A powered, 1974	100	150	___	___	___
8453	Erie Alco B dummy, 1974	75	100	___	___	___
8454	Rio Grande GP-7 powered, 1974-75	75	100	___	___	___
8455	Rio Grande GP-7 dummy, 1974-75	30	50	___	___	___
8456	Norfolk Southern GP-7, 1974	60	75	___	___	___
8458	Erie-Lack. SD-40 twin motors, 1985	325	400	___	___	___
8459	Rio Grande Rotary Snowplow, 1984	100	135	___	___	___
8460	M K T NW-2 pow. Switcher, 1973-75	50	75	___	___	___
8463	Chessie GP-20 powered, 1974	100	125	___	___	___
8464/8465	Rio Grande F-3 AA set, 1974	225	350	___	___	___
8466/8467	Amtrak F-3 AA , 1974-75	375	575	___	___	___
8468	B & O F-3 B dummy, 1974	80	115	___	___	___

		Exc	LN	Color	Cond	$
8469	C P F-3 B dummy, 1974	200	300			
8470	Chessie U36B powered, 1974	100	150			
8471	Penn NW-2 pow. Switcher, 1973-74	150	200			
8473	Coca-Cola NW-2 pow. Switcher, 75	125	175			
8474	Rio Grande F-3 dummy, 1975	125	175			
8475	Amtrak F-3 B dummy, 1975	80	115			
8477	N Y C GP-9 powered, 1985	200	275			
8480/8481/8482	U P F-3 ABA units, 1985	450	550			
8485	U. S. Marines NW-2 Switcher, 1984	75	100			
8500	Pennsylvania Steam, 2-4-2, 1976	20	25			
8501	(See 18501)					
8502	Santa Fe Steam, 2-4-0, 1975	20	25			
8506	Pennsylvania Steam, 0-4-0, 1975-77	100	150			
8507	A T S F Steam, 2-4-0, 1975	25	30			
8510	Pennsylvania Steam, 0-4-0, 1975	20	30			
8512	Santa Fe Steam, 0-4-0, blue, 1985-86	25	35			
8516	N Y C Steam, 0-4-0, 1985-86	125	150			
8550	Jersey Central GP-9 powered, 1975	80	95			
8551	Pennsylvania EP-5 powered, 1975	200	275			
8552/8553/8554	S P Alco ABA, 1975-76	250	350			
8555/8557	Milwaukee Rd. F-3 AA, 1975	375	475			
8556	Chessie NW-2 pow. Switcher, 1975-76	200	295			
8558	Milwaukee Road EP-5, pow., 1976	125	175			
8559	(See 1776)					
8560	Chessie U36B dummy, 1975	75	100			
8561	Jersey Central GP-9 dummy, 1975-76	50	60			
8562	MoPac GP-20 powered, 1975-76	70	85			
8563	Rock Island Alco A powered, 1975	75	100			
8564	Union Pacific U36B powered, 1975	100	125			
8565	MoPac GP-20 dummy, 1975-76	50	75			
8566	Southern F-3 A powered, 1975-77	250	400			
8567	Southern F-3 A dummy, 1975-77	125	200			
8568	Preamble Exp. F-3 A powered, 1975	100	150			
8569	SOO NW-2 pow. Switcher, 1975-77	75	100			
8570	Liberty Spec. Alco A powered, 1975	75	100			
8571	Frisco U36B powered, 1975-76	75	100			
8572	Frisco U36B dummy, 1975-76	40	50			
8573	Union Pacific U36B dummy, 1975	200	275			
8575	Milwaukee Road F-3 B dummy, 1975	100	175			
8576	Penn Cent. GP-7 powered, 1975-76	75	85			
8578	N Y C Ballast Tamper pow., 1985, 87	100	125			
8580/8582	Illinois Central F-3 AA, 1985-87	325	400			
8581	Ill. Central F-3 B with horn, 1985, 87	150	200			
8585	B N SD-40 twin motors, horn, 1985	275	325			
8587	Wabash, GP-9 (J. C. Penney Spec.), 85	225	300			
8600	N Y C Steam, 4-6-4, 1976	250	300			
8601	Rock Island Steam, 0-4-0, 1976-77	20	25			
8602	Rio Grande Steam, 2-4-0, 1976-78	25	30			
8602	(See 18602)					
8603	C & O Steam, 4-6-4, 1976-77	200	225			
8604	Jersey Central 2-4-2, Steam, 1976	40	45			
8604	(See 18604)					
8606	Boston & Albany 773 Hudson, 1986	1400	1750			
8606	(See 18606)					
8607	(See 18607)					

		Exc	LN	Color	Cond	$
8610	Wabash 4-6-2 Steam, 1986-87	500	650			
8612	(See 18612)					
8615	Berkshire 2-8-4 Steam, J. C. Penney Catalog, 1986	750	100			
8616	A T & S F Steam 4-4-2, 1986	65	75			
8617	N K P Steam 4-4-2, 1986	65	75			
8625	Penn. Steam 2-4-0, 1986	25	35			
8630	W & A Steam 4-4-0, General, 1986	—	135			
8635	A T & S F Steam Switcher 0-4-0, 86	120	145			
8650	B N U36B powered, 1976-77	85	110			
8651	B N U36B dummy, 1976-77	55	75			
8652	Santa Fe F-3 A powered, 1976-77	200	300			
8653	Santa Fe F-3 A dummy, 1976-77	225	350			
8654	Boston & Maine GP-9 powered, 1976	75	110			
8655	Boston & Maine GP-9 dummy, 1976	30	50			
8656/8657/8658	C N Alco ABA units, 1976	200	300			
8656	C N Alco A unit powered, 1976	75	150			
8657	C N Alco B dummy unit only, 1976	50	75			
8658	C N Alco A dummy unit only, 1976	100	200			
8659	Virginian Rectifier, 1976-77	125	175			
8660	C P Rail NW-2 Switcher, 1976-77	60	95			
8661	Southern F-3 B dummy	75	100			
8662	B & O GP-7, 1986	75	100			
8664	Amtrak Alco A powered, 1976-77	75	100			
8665	(See 1776)					
8666	Northern Pacific GP-9 powered, 1976	100	125			
8667	Amtrak Alco B dummy, 1976-77	75	100			
8668	N P GP-9 dummy, 1976	50	75			
8669	Illinois Central U36B powered, 1976	100	125			
8670	Chessie Gas Turbine, 1976	35	65			
8679	Northern Pacific GP-20, 1986	80	100			
8687	Jersey Central FM Diesel, 1986	300	375			
8690	Lionel Lines Trolley Car, 1986	100	110			
8700	(See 18700)					
8701	W & A R R (Gen.) 4-4-0 "#3", 77-78	125	175			
8702	Crescent Limited Steam, 4-6-4, 1977	300	425			
8702	(See 18702)					
8703	Wabash Steam, 2-4-2, 1977	25	35			
8750	The Rock GP-7 powered, 1977	70	95			
8751	The Rock GP-7 dummy, 1977	30	40			
8753	Penn GG-1, tuscan, 1977	450	550			
8754	New Haven Rectifier, 1977-78	125	175			
8755	Santa Fe U36B powered, 1977-78	85	100			
8756	Santa Fe U36B dummy, 1977-78	50	70			
8757	Conrail GP-9 powered, 1977-78	75	90			
8758	Southern GP-7 dummy, 1978	40	60			
8759	Erie-Lack. GP-9 powered, 1977-79	90	125			
8760	Erie-Lack. GP-9 dummy, 1977-79	50	80			
8761	G T W NW-2 pow. Switcher, 1977-78	200	295			
8762	Great Northern EP-5, 1977-78	200	275			
8763	N & W GP-9 powered, 1977-78	75	95			
8764	B & O Budd RDC powered, 1977	100	150			
8765	B & O Budd RDC dummy, 1977	75	100			
8766	B & O Budd RDC powered with 8767, 8768, 1977	300	350			

		Exc	LN	Color	Cond	$
8767	B & O RDC dummy, 1977 (See 8766)					
8768	B & O RDC dummy, 1977 (See 8766)					
8769	Republic Steel Gas Turbine, 1977	20	40	___	___	___
8770	E M D NW-2 pow. Switcher, 1977	60	95	___	___	___
8771	Great Northern U36B powered, 1977	100	120	___	___	___
8772	G M & O GP-20 powered, 1977	70	100	___	___	___
8773	Mickey Mouse U36B pow., 1977-78	300	500	___	___	___
8774	Southern GP-7 powered, 1977-78	80	100	___	___	___
8775	Lehigh Valley GP-9 pow., 1977-78	85	100	___	___	___
8776	C & N W GP-20 powered, 1977-78	85	100	___	___	___
8777	Santa Fe F-3 B unit dummy, 1977-78	200	300	___	___	___
8778	Lehigh Valley GP-9 dummy, 1977-78	60	90	___	___	___
8779	C & N W GP-20 dummy, 1977-78	60	85	___	___	___
8800	(See 18800)					
8800	Lionel Lines Steam, 4-4-2, 1978-81	100	125	___	___	___
8801	(See 18801)					
8801	Blue Comet Steam, 4-6-4, 1979-80	350	425	___	___	___
8802	(See 18802)					
8803	Santa Fe Steam, 0-4-0, 1979	10	20	___	___	___
8806	(See 18806)					
8850	Penn Central GG-1, black, 1979	375	475	___	___	___
8851/8852	New Haven F-3, AA, pr., 78-79	300	400	___	___	___
8854	C P Rail GP-9 powered, 1978-79	85	125	___	___	___
8855	Milwaukee Road SD-18 pow., 1978	115	150	___	___	___
8857	Northern Pacific U36B pow., 1978-80	70	100	___	___	___
8858	N P U36B dummy, 1978-80	30	40	___	___	___
8859	Conrail Rectifier, 1978-80, 1982	100	150	___	___	___
8860	Rock NW-2 pow. Switcher, 1978-79	75	95	___	___	___
8861	Santa Fe Alco A powered, 1978-79	75	100	___	___	___
8862	Santa Fe Alco A dummy, 1978-79	40	50	___	___	___
8864	New Haven F-3 B dummy, 1978	125	175	___	___	___
8866	M & St L GP-9 powered, 1978	85	95	___	___	___
8867	M & St L GP-9 dummy, 1978	40	50	___	___	___
8868	Amtrak Budd RDC pow., 1978, 1980	150	200	___	___	___
8869	Amtrak Budd RDC dum., 1978, 1980	70	100	___	___	___
8870	Amtrak Budd RDC dum., 1978, 1980	70	100	___	___	___
8871	Amtrak Budd RDC dum., 1978, 1980	70	100	___	___	___
8872	Santa Fe SD-18 powered, 1978-79	100	150	___	___	___
8873	Santa Fe SD-18 dummy, 1978-79	50	85	___	___	___
8900	A T & S F Steam, 4-6-4, 1979	350	425	___	___	___
8900	Diesel Switcher, "8900", 1989	NRS	NRS	___	___	___
8902	A C L Steam, 2-4-0, 1979-82	15	20	___	___	___
8903	Rio Grande Steam, 2-4-2, 1979	20	25	___	___	___
8904	Wabash Steam, 2-4-2, 1979, 1981	35	40	___	___	___
8905	Steam (no letters), 0-4-0, 1979	10	20	___	___	___
8950	Virginian Trainmaster, 1978	350	495	___	___	___
8951	S P Trainmaster, 1979	475	600	___	___	___
8952/8953	Penn. F-3, AA, green, 1979	650	850	___	___	___
8955	Southern U36B powered, 1979	100	175	___	___	___
8956	Southern U36B dummy, 1979	50	75	___	___	___
8957	B N GP-20 powered, green, 1979	80	110	___	___	___
8958	B N GP-20 dummy, green, 1979	40	60	___	___	___
8960	S P U36C, pair, 1979	80	100	___	___	___
8961	S P U36C dummy, 1979	40	50	___	___	___
8970/8971	Penn. F-3, AA, pair, 1979-80	400	500	___	___	___

		Exc	LN	Color	Cond	$
8962	Reading U36B, 1979	75	100	___	___	___
8977	(See 18000)					
9001	Conrail Boxcar, 1986	12	15	___	___	___
9010	Great Northern Hopper, blue, 1971	6	8	___	___	___
9011	Great Northern Hopper, 1971, 1979	6	8	___	___	___
9012	T A & G Short Hopper, 1971-72, 1979	5	7	___	___	___
9013	C N Hopper, red, 1972-74, 1979	4	6	___	___	___
9014	Trailer Train Flatcar (O27), 1978	4	6	___	___	___
9015	Reading Short Hopper, 1973-74, 1979	20	25	___	___	___
9016	Chessie Short Hop., yellow, 1975-79	4	6	___	___	___
9016	Chessie Short Hop., "LCCA 1979-80 MEET SPECIAL"1981	—	25	___	___	___
9017	C P Short Hopper, 1971	6	8	___	___	___
9017	Wabash Gondola (O27), 1978, 1980-81	4	5	___	___	___
9018	D T & I Short Hopper, 1978, 1981	5	8	___	___	___
9019	Unlettered Set Flatcar (O27), 1978	1	2	___	___	___
9020	Union Pacific Flatcar (O27), 1970-77	4	5	___	___	___
9021	Santa Fe Work Caboose, 1970-74	7	10	___	___	___
9022	A T & S F Bulkhead Flatcar, 1971, 1977	8	15	___	___	___
9023	M K T Bulkhead Flatcar, 1974, 1978	8	15	___	___	___
9024	C & O Flatcar, yellow (O27), 1974	3	4	___	___	___
9025	D T & I Short Hopper, 1978	5	7	___	___	___
9025	D T & I Work Caboose, 1971-74, 1978	8	10	___	___	___
9025	D T & I Flatcar, 1978	NRS	NRS	___	___	___
9026	Republic Steel Flatcar, 1975-77, 1980	3	4	___	___	___
9027	SOO Work Caboose, 1975	8	10	___	___	___
9028	B & O Short Hopper, 1978	NRS	NRS	___	___	___
9030	Kickapoo, 1972, 1979	5	7	___	___	___
9031	Nickel Plate Gondola, 1974, 1979, 1983	5	7	___	___	___
9032	S P Gondola (O27), 1975, 1978	3	4	___	___	___
9033	Penn Central Gon., 1977, 1979, 1981-82	3	4	___	___	___
9034	Lionel Leisure Short Hopper (O27), 77	35	50	___	___	___
9035	Conrail Boxcar (O27), 1978-82	4	8	___	___	___
9036	Mobilgas 1-D Tank, white, 1978-80	10	15	___	___	___
9036	Mobilgas Tank Car, L C C A, 1978-79	—	35	___	___	___
9037	Conrail Boxcar (O27), 1978-81	4	8	___	___	___
9038	Chessie Short Hopper, blue, 1975-79	8	10	___	___	___
9039	Cheerios Boxcar (O27), 1971-72	6	8	___	___	___
9039	Mobilgas 1-D Tank, red, 1978, 1980	10	15	___	___	___
9040	Wheaties Boxcar (O27), 1970	7	10	___	___	___
9041	Hershey's Boxcar (O27), 1971	7	10	___	___	___
9042	Autolite Boxcar (O27), 1972	8	12	___	___	___
9043	Erie-Lack. Boxcar (O27), 1973-74	8	10	___	___	___
9044	D & R G W Boxcar (O27), 1975, 1979	6	8	___	___	___
9045	Toys 'R Us Boxcar (O27)	40	50	___	___	___
9046	True Value Boxcar (O27), 1976	40	50	___	___	___
9047	Toys 'R Us Boxcar (O27)	35	50	___	___	___
9048	Toys 'R Us Boxcar (O27)	35	50	___	___	___
(9049)	Toys 'R Us Boxcar, 1979	35	50	___	___	___
9050	Sunoco 1-D Short Tank, 1970-71	10	15	___	___	___
9051	Firestone 1-D Tank, white, 1974-75	10	15	___	___	___
9052	Toys 'R Us Short Boxcar, 1977	30	40	___	___	___
9053	True Value Short Boxcar	35	50	___	___	___
9054	J C Penney Short Boxcar	35	50	___	___	___
9055	Republic Steel Gondola (O27), 1977-81	10	12	___	___	___

		Exc	LN	Color	Cond	$
9057	C P Rail SP Caboose, 1978-79	9	11			
9058	Lionel Lines SP Caboose, 1978-79	6	8			
9060	Nickel Plate SP Caboose, 1970-71	6	8			
9061	A T & S F SP Caboose, 1970-71, 1978	6	8			
9062	Penn Central SP Caboose, 1970-71	5	9			
9063	G T W SP Caboose, 1970	20	25			
9064	C & O SP Caboose, 1971	7	10			
9065	C N SP Caboose, 1971-72	20	25			
9066	Southern SP Caboose	8	10			
9067	Kickapoo Valley Bobber Caboose, 1972	7	10			
9068	Reading Bobber Caboose, 1973-75	6	8			
9069	Jersey Central SP Caboose, 1973-74	6	8			
9070	Rock Island SP Caboose, 1973-74	10	15			
9071	A T & S F Bobber Caboose, 1974-75	8	10			
9073	Coca-Cola SP Caboose, 1973	10	20			
9075	Rock Island SP Caboose, 1973-74	10	15			
9076	We The People SP Caboose, 1975	15	25			
9077	Rio Grande SP Caboose, 1977-79, 1981	6	8			
9078	Rock Island Bobber Cab., 1977-79	6	8			
9079	A T & S F Caboose, 1979			Not Manufactured		
9079	G T W Short Hopper, 1977	9	12			
9080	Wabash SP Caboose, 1977	10	11			
9085	A T & S F Work Caboose, 1980-81	5	6			
9090	Mini-Max four-wheel Van Car, 1971	30	40			
9100	(See 19100)					
9100	(See 18205)					
9101	(See 19101)					
9102	(See 19102)					
9103	(See 19103)					
9104	(See 19104)					
9106	(See 19106)					
9106	Easy Cheese Vat Car, 1984-85	NRS	NRS			
9106	Miller Vat Car, 1984	15	20			
9107	Dr. Pepper Vat Carrier, 1986-87	15	20			
9110	B & O Quad Hopper, 1971	25	40			
9111	N & W Quad Hopper, 1972	15	20			
9112	D & R G Quad Hopper, 1972-73	15	20			
9113	N & W Quad Hopper, 1973	25	40			
9114	Morton's Quad Hopper, 1975-76	15	25			
9115	Planter's Quad Hopper, 1974-76	15	25			
9116	Domino Sugar Quad Hopper, 1974-76	15	25			
9117	Alaska Quad Hopper, 1974-76	15	25			
9118	L C C A Corning Quad Hopper, 1974	75	100			
9119	Detroit & Mack. Quad Hopper, 1975	35	45			
9120	Northern Pac. Flatcar w/vans, 1970-71	40	60			
9121	L & N Flatcar w/dozer, 1974, 76, 78-79	35	55			
9122	N P Flatcar w/ vans, 1972-75	40	60			
9123	Chesapeake & Ohio Auto Carrier, 1974	15	25			
9124	P & L E Flatcar with logs, 1973	8	15			
9125	N & W Auto Carrier, 1974	25	40			
9126	C & O Auto Carrier, 1973-74	15	25			
9128	Heinz Vat Car, 1974-76	15	20			
9129	N & W Auto Carrier, brown, 1975	20	30			
9130	B & O Quad Hopper, 1970-71	15	25			
9131	Rio Grande Gondola, orange, 1974	4	8			

		Exc	LN	Color	Cond	$
9132	Libby's Vat Car, 1975-77	15	20	___	___	___
9133	B N Flatcar with van, green, 1976, 80	40	60	___	___	___
9134	Virginian Quad Hopper, 1976-77	25	30	___	___	___
9135	N & W Quad Hopper, blue, 1971	15	20	___	___	___
9136	Republic Steel Gondola, blue, 1976-79	6	8	___	___	___
9138	Sunoco 3-D Tank, black, 1978	30	40	___	___	___
9139	Penn Central Auto Carrier, green, 1977	20	30	___	___	___
9140	Burlington Gondola, 1970-71, 1980-81	5	10	___	___	___
9141	B N Gondola 1970-71, 1980, 1981	3	4	___	___	___
9142	Republic Steel Gondola, green, 1971	3	4	___	___	___
9142	Republic Steel Gondola, "1977-1978 MEET SPECIAL", 1971	—	25	___	___	___
9143	Canadian National Gondola, 1973	20	25	___	___	___
9144	Rio Grande Gondola, black, 1974	4	5	___	___	___
9145	I C Gulf Auto Carrier, orange, 1977	20	35	___	___	___
9146	Mogen David Vat Car, 1977-79	15	20	___	___	___
9147	Texaco 1-D Tank, chrome, 1977	20	40	___	___	___
9148	Dupont 3-D Tank, 1977-79, 1981	15	25	___	___	___
9149	C P Rail Flatcar with vans, 1977	30	50	___	___	___
9150	Gulf 1-D Tank Car, white, 1970-71	25	30	___	___	___
9151	Shell 1-D Tank Car, yellow, 1972	20	25	___	___	___
9152	Shell 1-D Tank Car, 1973-74	15	20	___	___	___
9153	Chevron 1-D Tank Car, 1974-76	15	30	___	___	___
9154	Borden 1-D Tank Car, chrome, 1975-76	20	35	___	___	___
9155	L C C A, Monsanto 1-D Tank Car, 1977	65	75	___	___	___
9156	Mobilgas 1-D Tank Car, 1976-77	20	30	___	___	___
9157	C & O Flatcar w/crane, 1976-78, 1981	50	75	___	___	___
9158	P C Flatcar w/shovel, 1976-77, 1980	50	75	___	___	___
9159	Sunoco 1-D Tank Car, silver, 1975	35	50	___	___	___
9160	(See 8030)					
9160	Illinois Central N5C Caboose, 1970-72	20	35	___	___	___
9161	C N N5C Caboose, 1971-72	20	35	___	___	___
9162	Pennsylvania N5C Caboose, 1972-76	20	35	___	___	___
9163	A T & S F N5C Caboose, 1973-76	15	25	___	___	___
9165	Canadian Pacific N5C Caboose, 1973	25	40	___	___	___
9166	Rio Grande SP Caboose, 1974	12	20	___	___	___
9167	Chessie N5C Caboose, 1974-76	20	35	___	___	___
9168	Union Pacific N5C Caboose, 1975-76	15	30	___	___	___
9169	Milwaukee Road SP Caboose, 1975	15	25	___	___	___
9170	N & W (See 1776, 1976)					
9171	MoPac SP Caboose, 1975-77	15	20	___	___	___
9172	Penn Central SP Caboose, 1975-77	15	25	___	___	___
9173	Jersey Central SP Caboose, 1975-77	15	20	___	___	___
9174	P & L E B/W Caboose, green, 1976	75	90	___	___	___
9175	Virginian N5C Caboose, 1975-77	20	30	___	___	___
9176	B A R N5C Caboose, 1976	20	35	___	___	___
9177	N P B/W Caboose, green, 1976	20	30	___	___	___
9178	Illinois Central SP Caboose	15	25	___	___	___
9179	Chessie Bobber Caboose, 1979	6	10	___	___	___
9180	The Rock N5C Caboose, 1977-1978	15	25	___	___	___
9181	B & M N5C Caboose, blue, 1977	15	25	___	___	___
9182	N & W N5C Caboose, black, 1977-80	15	30	___	___	___
9183	Mickey Mouse N5C Caboose, 1977-78	40	60	___	___	___
9184	Erie Bay Window Caboose, 1977-78	20	30	___	___	___
9185	G T W N5C Caboose, 1977	15	30	___	___	___

		Exc	LN	Color	Cond	$
9186	Conrail N5C Caboose, 1977-78	15	25			
9187	G M & O SP Caboose, 1977-78	15	30			
9188	Great Northern B/W Cab., blue, 1977	30	45			
9189	Gulf 1-D Tank Car, chrome, 1977	35	50			
9189	Norfolk Southern NC5 Caboose, 1974	12	15			
9193	Budweiser Vat Car, 1983	15	20			
9195	Rolling Stock Asst. (O27),1988	NA	NA			
9200	Illinois Central Boxcar, 1970-72	25	35			
9201	Penn Central Boxcar, 1970	25	35			
9202	Santa Fe Boxcar, red, 1970	30	50			
9203	Union Pacific Boxcar, 1970	35	50			
9204	Northern Pacific Boxcar, green, 1970	30	45			
9205	Norfolk & Western Boxcar, 1970	20	25			
9206	Great Northern Boxcar, blue, 1970-71	20	25			
9207	SOO Boxcar, red, 1971	20	25			
9208	C P Rail Boxcar, yellow, 1971-72	20	25			
9209	B N Boxcar, green, 1970-72	20	25			
9210	B & O Automobile Boxcar, 1971	20	30			
9211	Penn. Central Boxcar, 1971	30	35			
9212	L C C A Flatcar w/vans, Atlanta, 1976	35	40			
9213	M & St L Quad Hopper, 1978	25	35			
9214	Northern Pacific Box., tuscan, 1970-72	25	35			
9215	Norfolk & Western Boxcar, 1971	30	35			
9216	Great Northern Auto Carrier, 1978	25	40			
9217	SOO Lines Oper. Box, 1982-83, 1985	15	25			
9218	Monon Operating Boxcar, 1981-82	15	25			
9219	M P Operating Boxcar, 1983	25	35			
9220	Borden Oper. Milk Car Set, 1985-86	70	85			
9221	Poultry Dispatch, 1985-86	30	35			
9222	L & N Flatcar with vans, 1983	15	20			
9223	Reading Operating Boxcar, 1985	25	30			
9224	Louisville Oper. Horse Car Set, 1985	75	90			
9225	Conrail Operating Barrel Car, 1984-85	35	50			
9226	D & H Flatcar with vans, 1985	—	15			
9227	Canadian Pacific Boxcar, 1986	20	25			
9228	Canadian Pacific Boxcar, 1986	20	30			
9229	Express Mail Oper. Boxcar, 1985-86	25	35			
9230	Monon Boxcar, brown, 1972	15	25			
9231	Reading B/W Caboose, 1979	25	30			
9232	Allis Chalmers Flatcar w/load, 1980	25	40			
9233	Die-cast Transformer Flatcar, 1980	60	85			
9234	Radioactive Waste Car, 1980	30	40			
9235	Union Pacific Flatcar with derrick, 1983	10	15			
9236	C & N W Derrick Car, 1985	15	20			
9237	U P S Operating Boxcar, 1984			Not Manufactured		
9238	N P Operating Log Dump, 1985	15	20			
9239	Lionel Lines N5C Caboose, 1985	35	50			
9240	N Y C Hopper Car, 1986	25	35			
9241	Penn Oper. Log Dump Car, 1985-86	15	20			
9245	Illinois Central Derrick Car, 1985			Not Manufactured		
9247	N Y C Searchlight Car (See 6529)			Not Manufactured		
9250	Waterpoxy 3-D Tank Car, 1971	20	25			
X9259	Southern Bay Window Cab., 1977	35	40			
9260	Reynolds Quad Hopper, 1975-78	15	20			
9261	Sunmaid Quad Hopper, 1975-76	15	20			

		Exc	LN	Color	Cond	$
9262	Ralston-Purina Quad Hopper, 1975-76	75	90			
9263	Pennsylvania Quad Hopper, 1975-77	25	30			
9264	Ill. Central Quad Hopper, 1975-77	20	25			
9264	Ill. Central Hopper, "T C A MUSEUM EXPRESS, MARCH 8, 1980", 1980	—	100			
9265	W M Chessie Quad Hopper, 1975-77	25	30			
9266	Southern "Big John" Quad, 1976	40	50			
9267	Alcoa Quad Hopper, 1975	30	40			
9268	N P B/W Caboose, black, 1977-78	30	40			
9269	Milwaukee Road B/W Caboose, 1978	35	60			
9270	N P N5C Caboose, 1978, 1980	15	25			
9271	M & St L B/W Caboose, 1978-79	20	30			
9272	New Haven B/W Caboose, 1978-80	15	25			
9273	Southern Bay Window Cab., 1978	25	40			
9274	Santa Fe Bay Window, red, 1978	60	90			
9275	Santa Fe Bay Window, red, 1978	NRS	NRS			
9276	Peabody Open Quad Hopper, 1978	25	35			
9276	T & P SP Caboose, 1980			Not Manufactured		
9277	Cities Service 3-D Tank Car, 1977	30	40			
9278	Lifesavers 1-D Tank Car, 1978-79	30	50			
9279	Magnolia 3-D Tank Car, 1978-79	15	20			
9280	A T & S F Horse Transport, 1978-80	10	20			
9281	A T S F 2-level carrier, 1978-79	20	30			
9282	G N Flatcar with vans, 1978, 1981-82	40	60			
9283	Union Pacific Gondola, 1977	8	10			
9284	A T & S F Gondola, red, 1977-78	12	15			
9285	I C Gulf Flatcar w/vans, 1977	60	75			
9286	B & L E Quad Hopper, 1977	15	25			
9287	Southern N5C Caboose, 1978	15	25			
9288	Lehigh Valley N5C Cab., 1978, 1980	15	20			
9289	C & N W N5C Caboose, 1978, 1980	20	30			
9289	C & N W "T C A MUSEUM EXPRESS, MARCH 8, 1980", 1980	—	100			
9290	U P Operating Barrel Car, 1983	55	85			
9300	P C Dump Car, green, 1970-73	15	20			
9301	Operating U. S. Mail Car, 1975-83	15	30			
9302	L & N Searchlight Car, 1973-74	12	15			
9303	U P Operating Log Dump, 1974, 1979	10	15			
9304	C & O Coal Operating Dump, 1973-76	15	20			
9305	Santa Fe Animated Stock, 1980, 1982	15	20			
9306	A T & S F Animated Stock Car, 1980	15	20			
9306	C & O Oper. Coal Dump Car, 1974-76	10	15			
9307	Erie Animated Gondola, 1979-83, 1985	35	55			
9308	Aquarium Car, 1981-83, 1985	60	75			
9309	T P & W B/W Caboose, 1980-81	20	35			
9310	A T & S F Log Dump, 1978-79, 81-82	10	15			
9311	U P Operating Coal Dump, 1978-82	10	15			
9312	Conrail Searchlight Car, 1978-83	15	20			
9313	Gulf 3-D Tank Car, black, 1979	35	50			
9315	Southern Pacific Gondola, 1979	20	30			
9316	Southern Pacific B/W Caboose, 1979	35	65			
9317	A T & S F B/W Caboose, blue, 1979	25	35			
9319	T C A Silver Jubilee Gold Bullion Car, 1979	250	325			
9320	Fort Knox Gold Bullion Car, 1979	175	250			

		Exc	LN	Color	Cond	$
9321	A T & S F 1-D Tank (FARR #1), 1979	30	45	___	___	___
9322	A T & S F Quad Hopper (FARR #1), 79	70	80	___	___	___
9323	A T & S F B/W Caboose, brown, 1979	30	45	___	___	___
9324	Tootsie Roll 1-D Tank, 1979, 1981-82	25	35	___	___	___
9325	(See 9363)					
9325	N & W Flatcar w/ fence, black, 1980-81	5	10	___	___	___
9325	N & W Flatcar with cab, red, 1978-79	5	10	___	___	___
9325	U S M C Security Caboose, 1983	5	10	___	___	___
9326	B N B/W Caboose, green, 1979-80	20	25	___	___	___
9327	Bakelite 1-D Tank Car, 1980	15	20	___	___	___
9328	W M Chessie B/W Caboose, 1980	30	50	___	___	___
9329	W M Chessie Crane, 1980	45	70	___	___	___
9330	Kickapoo Valley Dump Car, 1972	3	5	___	___	___
9331	Union 76 1-D Tank Car, 1979	40	60	___	___	___
9332	Reading B/W Caboose, 1979	35	55	___	___	___
9332	Reading Crane, 1979	40	55	___	___	___
9333	S P Flatcar with vans, 1980	40	50	___	___	___
9334	Humble 1-D Tank Car, 1979	20	30	___	___	___
9335	B & O Operating Log Dump, 1986	—	20	___	___	___
9336	C P Rail Gondola, 1979	20	25	___	___	___
9338	Penn Power Quad Hopper, 1979	40	65	___	___	___
9339	G N Boxcar (O27), 1979-81, 1983	5	10	___	___	___
9340	Illinois Central Gondola (O27), 1979-81	5	10	___	___	___
9341	Atlantic Coast Line SP Cab., 1979-82	6	10	___	___	___
9344	Citgo 3-D Tank, 1980	35	45	___	___	___
9345	Reading Searchlight Car, 1985	20	25	___	___	___
9346	N Y N H & H SP Caboose, 1980-81	6	8	___	___	___
9346	Wabash SP Caboose, 1979	6	10	___	___	___
9347	Niagara Falls T T O S 1-D Tank, 1979	40	50	___	___	___
9348	Santa Fe Crane, 1979	55	70	___	___	___
9349	San Francisco Mint, 1980	110	150	___	___	___
9351	Pennsylvania Auto Carrier, 1980	20	40	___	___	___
9352	C & N W Flatcar with vans, 1980	50	75	___	___	___
9353	Crystal 3-D Tank Car, 1980	15	25	___	___	___
9354	Pennzoil 1-D Tank Car, 1981	20	35	___	___	___
9355	D & H Bay Window Caboose, 1980	20	25	___	___	___
9356	Life Savers Stik-O-Pep Tank Car, 1980			Not Manufactured		
9357	Smokey Mountain Bobber Cab., 1979	5	10	___	___	___
9358	L C C A Sands of Iowa Quad Hop., 80	30	40	___	___	___
9359	National Basketball Boxcar, 1980	20	25	___	___	___
9360	National Hockey League Boxcar, 1980	20	25	___	___	___
9361	C & N W Bay Window Caboose, 1980	20	40	___	___	___
9362	Major League Baseball Boxcar, 1980	20	25	___	___	___
9363	N & W Operating Dump, 1979	25	30	___	___	___
9364	N & W Crane Car, 1978	8	10	___	___	___
9365	Toys 'R Us Boxcar (O27), 1979	40	50	___	___	___
9366	U P Quad Hopper (FARR #2), 1980	25	35	___	___	___
9367	U P 1-D Tank Car (FARR #2), 1980	35	45	___	___	___
9368	U P B/W Caboose (FARR #2), 1980	30	40	___	___	___
9369	Sinclair 1-D Tank Car, 1980	60	75	___	___	___
9370	Seaboard Gondola, 1980	20	25	___	___	___
9371	Lantic Sugar Quad Hopper, 1980	25	35	___	___	___
9372	Seaboard Bay Window Cab., 1980	30	40	___	___	___
9373	Getty 1-D Tank, white, 1980-81	20	25	___	___	___
9374	Reading Quad Hopper, 1980-81	40	50	___	___	___

		Exc	LN	Color	Cond	$
9376	T & P SP Caboose, 1981	20	25			
9378	Lionel Flatcar w/yellow derrick, 1981	15	20			
9379	A T & S F Gondola, black, 1980	20	25			
9379	Lionel Flatcar w/derrick, 1980			Not Manufactured		
9380	N Y N H & H SP Caboose, 1980	10	12			
9381	Chessie SP Caboose, 1980	8	10			
9382	Florida East Coast B/W Cab., 1980	15	20			
9383	U P Flatcar w/vans (FARR #2), 1980	30	40			
9384	G N Operating Short Hopper, 1981	55	75			
9385	Alaska Gondola, yellow, 1981	20	25			
9386	Pure Oil 1-D Tank Car, 1981	30	50			
9387	Burlington B/W Caboose, red, 1981	35	45			
9388	Toys 'R Us Boxcar (O27), 1981	40	50			
9389	Radioactive Waste Car, 1981	20	30			
9398	Penn Operating Coal Dump, 1983	15	20			
9399	C & N W Coal Dump, 1983-84	15	20			
9400	Conrail Boxcar, brown, 1978	12	15			
9401	Great Northern Boxcar, 1979	10	15			
9402	Susquehanna Boxcar, 1978	20	25			
9403	S C L Boxcar, black, 1978	12	15			
9404	Nickel Plate Boxcar, 1978-79	30	40			
9405	Chattahoochie Boxcar, 1978-79	10	12			
9406	D & R G W Boxcar, 1978-79	10	12			
9407	Union Pacific Cattle Car, 1978	15	20			
9408	Lionel Lines Circus Car, 1978	30	40			
9411	Lackawanna "Snow" Boxcar, 1978	35	50			
9412	R F & P Boxcar, 1979	12	18			
9413	Napierville Jct. Boxcar, 1979-80	12	15			
9414	Cotton Belt Boxcar, 1980	8	10			
9415	P & W Boxcar, 1979	10	12			
9416	M D & W Boxcar, 1979, 1981	8	10			
9417	C P Rail Boxcar, gold/black, 1980	35	45			
9418	F A R R Boxcar, 1979	75	100			
9419	Union Pacific Boxcar (FARR #2), 1980	25	35			
9420	B & O "Sentinel" Boxcar, 1980	25	35			
9421	Maine Central Boxcar, 1980	12	15			
9422	E J & E Boxcar, 1980	15	25			
9423	N Y N H & H Boxcar, 1980	20	25			
9424	T P & W Boxcar, orange, 1980	15	20			
9425	British Columbia Auto Boxcar, 1980	20	25			
9426	Chesapeake & Ohio Boxcar, 1980	20	35			
9427	Bay Line Boxcar, 1980-81	10	15			
9428	T P & W Boxcar, green, 1980	35	50			
9429	The Early Years Boxcar, 1980	25	30			
9430	Standard Gauge Years Boxcar, 1980	25	30			
9431	The Prewar Years Boxcar, 1980	25	30			
9432	The Postwar Years Boxcar, 1980	85	125			
9433	The Golden Years Boxcar, 1980	85	125			
9434	Joshua Lionel Cowen Boxcar, 1980	70	100			
9435	L C C A, Central of Georgia, 1981	40	50			
9436	Burlington Boxcar, red, 1981	40	55			
9437	Northern Pacific Cattle Car, 1981	30	50			
9438	Ontario Northland Boxcar, 1981	20	25			
9439	Ashley Drew & Northern Boxcar, 1981	10	15			
9440	Reading Boxcar, 1981	40	70			

		Exc	LN	Color	Cond	$
9441	Pennsylvania Boxcar, 1981	40	70			
9442	Canadian Pacific Boxcar, 1981	15	20			
9443	F E C Boxcar, 1981	15	20			
9444	Louisiana Midland Boxcar, 1981	15	20			
9445	Vermont Northern Boxcar, 1981	15	20			
9446	Sabine River Boxcar, 1981	15	20			
9447	Pullman Standard Boxcar, 1981	15	20			
9448	A T & S F Cattle Car, brown, 1981	40	55			
9449	G N Boxcar, green (FARR #3), 1981	35	50			
9450	G N Cattle, red (FARR #3), 1981	50	80			
9451	Southern Boxcar (FARR #4), 1984	40	55			
9452	Western Pacific Boxcar, 1982-83	15	20			
9453	M P A Boxcar, 1982-83	10	15			
9454	New Hope & Ivyland Boxcar, 1982-83	10	15			
9455	Milwaukee Road Boxcar, 1982-83	15	20			
9456	P R R Auto Boxcar (FARR #5), 1985	35	50			
9460	D & T S Auto Boxcar, L C C A, 1982	35	60			
9461	Norfolk & Southern Boxcar, 1983	35	60			
9462	Southern Pacific Boxcar, 1983-84	20	25			
9463	Texas & Pacific Boxcar, 1983-84	12	15			
9464	N C & St L Boxcar, 1983-84	12	15			
9465	A T & S F Boxcar, green, 1983	12	25			
9466	Wanamaker Boxcar, 1982	115	150			
9467	Tenn. World's Fair Boxcar, 1982	35	45			
9468	Union Pacific Auto Boxcar, red, 1983	40	50			
9469	N Y C Pacemaker Box. (Std. O), 1985	85	125			
9470	Chicago Beltline Boxcar, 1984	15	20			
9471	Atlantic Coast Line Boxcar, 1984	15	20			
9472	Detroit & Mackinac Boxcar, 1985	20	25			
9473	Lehigh Valley Boxcar, 1984	15	20			
9474	Erie-Lackawanna Boxcar, 1985	35	45			
9475	D & H "I Love N. Y." Boxcar, 1984-85	25	40			
9476	Pennsylvania Boxcar, 1985	25	40			
9480	M N & S Boxcar, 1985-86	15	20			
9481	Seaboard System Boxcar, 1985-86	15	20			
9482	Norfolk Southern Boxcar, 1985-86	15	20			
9483	Manufacturers Railway Box., 1985-86	12	15			
9484	Lionel 85th Anniv. Boxcar, 1985	20	30			
9486	"I Love Michigan" Boxcar, 1986	15	20			
9491	Christmas Car, 1986	35	40			
9492	Lionel Lines Boxcar, 1986	30	50			
9500	Milwaukee Road Pullman, 1973	25	40			
9501	Milwaukee Road Pullman, 1973	25	40			
9502	Milwaukee Road Observation, 1973	25	40			
9503	Milwaukee Road Pullman 1974	25	40			
9504	Milwaukee Road Pullman, 1974	25	40			
9505	Milwaukee Road Pullman, 1974	25	40			
9506	Milwaukee Road Baggage, 1975	25	40			
9507	Pennsylvania Pullman, 1974	30	50			
9508	Pennsylvania Pullman, 1974	30	50			
9509	Pennsylvania Observation, 1974	30	50			
9510	Pennsylvania Combine, 1975	30	50			
9511	Milwaukee Road Pullman, 1974	25	40			
9512	Summerdale Junction T T O S, 1974	40	50			
9513	Pennsylvania Pullman, 1975	30	50			

		Exc	LN	Color	Cond	$
9514	Pennsylvania Pullman, 1975	30	50			
9515	Pennsylvania Pullman, 1975	30	50			
9516	Baltimore & Ohio Pullman, 1975	25	40			
9517	Baltimore & Ohio Coach, 1975	30	40			
9518	Baltimore & Ohio Obs., 1975	30	40			
9519	Baltimore & Ohio Baggage, 1975	25	40			
9520	T T O S Coach, 1975	40	50			
9521	Pennsylvania Baggage, 1975	75	95			
9522	Milwaukee Road Baggage, 1975	60	75			
9523	Baltimore & Ohio Baggage, 1975	60	70			
9524	Baltimore & Ohio Pullman, 1976	30	50			
9525	Baltimore & Ohio Pullman, 1976	30	50			
9526	T T O S Coach, 1976	40	50			
9527	Milwaukee Campaign, 1976	30	50			
9528	Pennsylvania Campaign, 1976	30	50			
9529	Baltimore & Ohio Campaign, 1976	30	50			
9530	Southern Baggage, 1978	30	50			
9531	Southern Combine, 1978	30	50			
9532	Southern Pullman, 1978	30	50			
9533	Southern Pullman, 1978	30	50			
9534	Southern Observation, 1978	30	50			
9535	T T O S Observation, 1977	40	50			
9536	Blue Comet Baggage, 1978	30	50			
9537	Blue Comet Combine, 1978	30	50			
9538	Blue Comet Pullman, 1978	30	60			
9539	Blue Comet Pullman, 1978	30	60			
9540	Blue Comet Observation, 1978	30	50			
9541	Santa Fe Baggage, 1980, 1982	20	30			
9542	Baltimore & Ohio Pullman, 1976	30	50			
9544	T C A Chicago Pullman, 1980	—	75			
9545	U P Baggage, 1985	75	100			
9546	U P Combine, 1985	75	100			
9547	U P Observation, 1985	75	100			
9548	U P "Placid Bay" Coach, 1985	75	100			
9549	U P "Ocean Sunset" Coach, 1985	75	100			
9551	W & A Baggage, 1977-79	30	50			
9552	W & A Coach	30	50			
9553	W & A Flatcar with fences, 1978-79	20	40			
9554	Alton Limited Baggage, 1981	40	60			
9555	Alton Limited Combine, 1981	40	60			
9556	Alton Limited Coach, 1981	50	80			
9557	Alton Limited Coach, 1981	50	80			
9558	Alton Limited Observation, 1981	50	80			
9559	Rock Island Combine, 1981	30	40			
9560	Rock Island Coach, 1981	30	40			
9561	Rock Island Coach, 1981	30	40			
(9562)	N & W "577" Baggage, 1981	90	125			
(9563)	N & W "578" Combine, 1981	90	125			
(9564)	N & W "579" Coach, 1981	110	150			
(9565)	N & W "580" Coach, 1981	110	150			
(9566)	N & W "581" Observation, 1981	90	125			
(9567)	N & W "582" Vista Dome, 1981	350	450			
569	Pennsylvania Combine, 1981	125	150			
570	Pennsylvania Baggage, 1979	125	150			
571	Pennsylvania Pullman, 1979	140	175			

			Exc	LN	Color	Cond	$
	9572	Pennsylvania Pullman, 1979	140	175			
	9573	Pennsylvania Vista Dome, 1979	125	150			
	9574	Pennsylvania Observation, 1979	125	150			
	9575	Pennsylvania "Edison", 1979	135	175			
•	9576	Burlington Baggage, 1980	70	100			
•	9577	Burlington Coach, 1980	80	110			
•	9578	Burlington Coach, 1980	80	110			
•	9579	Burlington Vista Dome, 1980	80	110			
•	9580	Burlington Observation, 1980	80	110			
	9581	Chessie Baggage, 1980	60	80			
	9582	Chessie Combine, 1980	60	80			
	9583	Chessie Coach, 1980	60	80			
	9584	Chessie Coach, 1980	60	80			
	9585	Chessie Observation, 1980	60	80			
	9586	Chessie Special Dining Car, 1986	—	95			
•	9588	Burlington Vista Dome, 1980	110	150			
•	9589	Southern Pacific Baggage, 1982	100	125			
•	9590	Southern Pacific Combine 1982	100	125			
•	9591	Southern Pacific Pullman, 1982-83	100	125			
	9592	Southern Pacific Pullman, 1982-83	100	125			
	9593	Southern Pac. Observation, 1982-83	100	125			
	9594	N Y C Baggage, 1983	100	125			
	9595	N Y C Combine, 1983	115	150			
	9596	N Y C Pullman, 1983	115	150			
	9597	N Y C Pullman, 1983	115	150			
	9598	N Y C Observation, 1983	100	125			
	9599	Chicago & Alto Dining Car, 1986	—	90			
	9600	Chessie Hi-cube, 1976	15	25			
	9601	Illinois Central Hi-cube, 1976-77	20	25			
	9602	A T & S F Hi-cube, 1977	20	25			
	9603	Penn Central Hi-cube, 1976-77	20	25			
	9604	N & W Hi-cube, 1976-77	20	25			
	9605	New Haven Hi-cube, 1976-77	20	25			
	9606	Union Pacific High-cube, 1976-77	20	25			
	9607	Southern Pacific Hi-cube, 1976-77	20	25			
	9608	Burlington Northern Hi-cube, 1977	20	25			
	9610	Frisco Hi-cube, 1977	30	50			
	9611	T C A Hi-cube, 1978	35	50			
	9620	N H L Wales, 1980	15	20			
	9621	N H L Campbell, 1980	15	20			
	9622	N B A Western, 1980	15	20			
	9623	N B A Eastern, 1980	15	20			
	9624	National League, 1980	15	20			
	9625	American League, 1980	15	20			
	9626	A T & S F Hi-cube, 1982-83	10	15			
	9627	Union Pacific Hi-cube, 1982-83	10	15			
	9628	Burlington Northern Hi-cube, 1982-83	10	15			
	9629	Chessie System Hi-cube, 1983	15	20			
	9660	Mickey Mouse Hi-cube, 1977-78	20	35			
	9661	Goofy Hi-cube, 1977-78	20	35			
	9662	Donald Duck Hi-cube, 1977-78	20	35			
	9663	Dumbo Hi-cube, 1978	40	70			
	9664	Cinderella Hi-cube, 1978	30	50			
	9665	Peter Pan Hi-cube, 1978	30	50			
	9666	Pinocchio Hi-cube, 1978	115	150			

		Exc	LN	Color	Cond	$
9667	Snow White Hi-cube, 1978	250	350			
9668	Pluto Hi-cube, 1978	115	150			
9669	Bambi Hi-cube, 1978	50	75			
9670	Alice In Wonderland Hi-cube, 1978	50	75			
9671	Fantasia Hi-cube, 1978	30	40			
9672	Mickey Mouse "50th" Hi-cube, 1978	300	375			
9678	T T O S Hi-cube, 1978	40	50			
9700	Southern Boxcar, red, 1972-73	20	30			
9700-1976	(See 9779)					
9701	B & O Automobile Boxcar, 1971-73	30	40			
9702	SOO Boxcar, white, 1972-73	25	35			
9703	C P Rail Boxcar, red, 1970-71	55	65			
9704	N & W Boxcar, brown, 1972	15	25			
9705	D & R G W Boxcar, orange, 1972-73	15	25			
9706	C & O Boxcar, 1972-74	20	30			
9706	(See 19706)					
9707	M K T Cattle Car, red, 1972-74	20	30			
9708	U. S. Mail Boxcar, 1972-74	15	25			
9709	State of Maine Boxcar, 1973-74	30	40			
9710	Rutland Boxcar, 1973-74	30	40			
9711	Southern Boxcar, brown, 1974	20	25			
9712	B & O Automobile Boxcar, 1973-74	30	40			
9713	C P Rail Boxcar, green, 1973-74	20	25			
9714	D & R G W Boxcar, 1973-74	20	25			
9715	Chesapeake & Ohio Boxcar, 1973-74	20	25			
9716	Penn. Central Boxcar, 1973-74	20	30			
9717	Union Pacific Boxcar, 1973-74	20	30			
9718	Canadian National Boxcar, 1973-74	30	40			
9719	New Haven Automobile Boxcar	30	40			
9720	Assorted Case of Cars	NRS	NRS			
9721	Assorted Case of Cars	NRS	NRS			
9723	Western Pacific Boxcar, 1974	30	50			
9724	Missouri Pacific Boxcar, 1974	35	50			
9725	M K T Cattle Car, yellow, 1974-75	15	20			
9726	Erie-Lack. Boxcar, blue, 1978	25	35			
9727	T A G Boxcar, L C C A, 1973	200	250			
9728	U P L C C A Cattle Car, yellow, 1978	35	45			
9729	C P Rail Boxcar, black/white, 1979	40	60			
9730	C P Rail Boxcar, gray/white, 1974-75	25	30			
9730	C P Rail Boxcar, gray/black, 1974-75	20	25			
9731	Milwaukee Road Boxcar, 1974-75	20	25			
9732	Southern Pacific Boxcar, 1979	30	40			
9733	Airco L C C A Boxcar w/tank, 1979	40	55			
9734	Bangor & Aroostock Boxcar, 1979	35	45			
9735	G T W Boxcar, 1974-75	15	20			
9737	Central of Vermont Boxcar, 1974-75	20	25			
9738	Illinois Terminal Boxcar, 1982	55	70			
9739	D & R G W Boxcar, yellow, 1975	15	20			
9740	Chessie Boxcar, yellow, 1974-75	15	20			
9742	M & St L Boxcar, green, 1975	30	40			
9743	Sprite Boxcar, 1974	25	35			
9744	Tab Boxcar, 1974	25	35			
9745	Fanta Boxcar, 1974	25	35			
9747	Chessie System Boxcar, 1975-76	20	30			
9748	C P Rail Boxcar, blue, 1975-76	20	30			

		Exc	LN	Color	Cond	$
9749	Penn Central Boxcar, 1975	20	30			
9750	D T & I Boxcar, 1975-76	20	30			
9751	Frisco Boxcar, 1975-76	15	25			
9752	L & N Boxcar, 1975-76	15	25			
9753	Maine Central Boxcar, 1975-76	25	35			
9754	N Y C Pacemaker Boxcar, 1976-77	25	35			
9755	Union Pacific Boxcar, 1975-76	20	30			
9757	Central of Georgia Boxcar, 1974	30	40			
9758	Alaska Boxcar, 1976-77	30	40			
9759	Paul Revere Boxcar, 1975-76	40	55			
9760	Liberty Bell Boxcar, 1975-76	40	55			
9761	George Washington Boxcar, 1975-76	40	55			
9762	Welcome Toy Fair Boxcar, 1975	150	200			
9763	Rio Grande Cattle Car, 1976	25	35			
9764	G T W Auto Boxcar, 1976-77	20	25			
9767	Railbox Boxcar, 1976-77	20	25			
9768	Boston & Maine Boxcar, 1976-77	20	30			
9769	B & L E Boxcar, 1976-77	15	20			
9770	Northern Pacific Boxcar, 1976-77	15	20			
9771	Norfolk & Western Boxcar, 1976-77	30	40			
9772	Great Northern Boxcar, 1975	65	90			
9773	N Y C Cattle Car, yellow, 1976	25	35			
9774	Southern Belle T C A Boxcar, 1975	35	50			
9775	M & St L Boxcar, red, 1975	30	45			
9776	S P "Overnight" Boxcar, black, 1975	45	60			
9777	Virginian Boxcar, 1976-77	15	20			
9778	Season's Greetings Boxcar, 1975	150	200			
9779	T C A Philadelphia Boxcar, 1976	35	50			
9780	Johnny Cash Boxcar, 1976	30	40			
9781	Delaware & Hudson Boxcar, 1977-78	20	25			
9782	The Rock Boxcar, 1977-78	20	25			
9783	B & O "Time Saver" Boxcar, 1977-78	25	35			
9784	A T & S F Boxcar, red, 1977-78	25	35			
9785	Conrail Boxcar, blue, 1977-79	20	25			
9786	C & N W Boxcar, brown, 1977-79	25	35			
9787	Central of N J Boxcar, 1977-79	15	35			
9788	Lehigh Valley Boxcar, 1977-79	15	20			
9789	Pickens Boxcar, 1977	25	40			
9801	B & O Sentinel Boxcar (Std. O), 1975	30	50			
9802	Miller High Life Reefer (Std. O), 1975	30	40			
9803	Johnson's Wax Boxcar (Std. O), 1975	25	40			
9805	Grand Trunk Reefer (Std. O), 1975	30	40			
9806	Rock Island Boxcar (Std. O), 1975-76	75	110			
9807	Stroh's Beer Reefer (Std. O), 1975-76	80	150			
9808	Union Pacific Box. (Std. O), 1975-76	100	150			
9809	Clark Reefer (Std. O), 1975-76	40	50			
9811	Pacific Fruit Express Reefer, 1980	25	35			
9812	Arm & Hammer Reefer, 1980	15	20			
9813	Ruffles Reefer, 1980	15	20			
9814	Perrier Reefer, 1980	20	25			
9815	N Y C Reefer (Std. O), 1985	70	90			
9816	Brachs Reefer, 1980	15	20			
9817	Bazooka Gum Reefer, 1980	15	20			
9818	Western Maryland Reefer, 1980	25	35			
9819	Western Fruit Express Reefer, 1981	25	35			

		Exc	LN	Color	Cond	$
9820	Wabash Gon., black (Std. O), 1973-74	40	50	____	____	____
9821	S P Gondola, tuscan (Std. O), 1973-74	40	50	____	____	____
9822	G T W Gondola, blue (Std. O), 1974	40	50	____	____	____
9823	A T & S F Flat w/crates (Std. O), 1976	80	125	____	____	____
9824	N Y C Gondola, black (Std. O), 1975	60	80	____	____	____
9825	Schaefer Reefer (Std. O), 1976-77	60	80	____	____	____
9826	P & L E Boxcar (Std. O), 1976-77	90	125	____	____	____
9827	Cutty Sark Reefer, 1984	15	20	____	____	____
9828	J & B Reefer, 1984	15	20	____	____	____
9829	Dewars Reefer, 1984	15	20	____	____	____
9830	Johnny Walker Red Label Reefer, 1984	15	20	____	____	____
9831	Pepsi Cola Reefer, 1982	30	40	____	____	____
9832	Cheerios Reefer, 1982	15	25	____	____	____
9833	Vlasic Pickles Reefer, 1982	15	20	____	____	____
9834	Southern Comfort Reefer, 1983	15	20	____	____	____
9835	Jim Beam Reefer, 1983	15	20	____	____	____
9836	Old Grand-Dad Reefer, 1983	15	20	____	____	____
9837	Wild Turkey Reefer, 1983	15	20	____	____	____
9840	Fleischmann's Gin Reefer, 1985	15	20	____	____	____
9841	Calvert Gin Reefer, 1985	15	20	____	____	____
9842	Seagram's Gin Reefer, 1985	15	20	____	____	____
9843	Tanqueray Gin Reefer, 1985	15	20	____	____	____
9844	Sambuca Reefer, 1986	15	20	____	____	____
9845	Baileys Irish Cream Reefer, 1986	20	25	____	____	____
9846	Seagrams Vodka Reefer, 1986	15	20	____	____	____
9847	Wolfschmidt Vodka Reefer, 1986	15	20	____	____	____
9849	Lionel Reefer, orange, 1984	50	70	____	____	____
9850	Budweiser Reefer, 1973-76	12	20	____	____	____
9851	Schlitz Reefer, 1973-76	15	20	____	____	____
9852	Miller Reefer, 1973-76	15	20	____	____	____
9853	Cracker Jack Reefer (caramel), 1973	25	35	____	____	____
9853	Cracker Jack Reefer (white), 1974	15	20	____	____	____
9854	Baby Ruth Reefer, 1973-76	15	20	____	____	____
9855	Swift Reefer, 1974-76	20	30	____	____	____
9856	Old Milwaukee Reefer, 1974-76	15	20	____	____	____
9858	Butterfinger Reefer, 1973-76	15	25	____	____	____
9859	Pabst Reefer, 1974-75	15	20	____	____	____
9860	Gold Medal Reefer, 1973-76	15	20	____	____	____
9861	Tropicana Reefer, 1976-77	25	30	____	____	____
9862	Hamm's Reefer, 1975-76	20	25	____	____	____
9863	R E A Reefer, 1975-76	30	40	____	____	____
9864	T C A Seattle Reefer, 1974	60	75	____	____	____
9866	Coors Reefer, 1977	15	25	____	____	____
9867	Hershey's Reefer, 1976-77	20	30	____	____	____
9868	T T O S Reefer, 1980	60	75	____	____	____
9869	Santa Fe Reefer, white, 1975	35	50	____	____	____
9870	Old Dutch Cleanser Reefer, 1977-78	15	20	____	____	____
9871	Carling's Reefer, 1977-78, 1980	15	20	____	____	____
9872	Pacific Fruit Express Reefer, 1977-78	15	20	____	____	____
9873	Ralston Purina Reefer, 1978	15	20	____	____	____
9874	Miller Lite Beer Reefer, 1978-79	15	20	____	____	____
9875	A & P Reefer, 1979	15	20	____	____	____
9876	Central Vermont Reefer, 1978	30	40	____	____	____
9877	Gerber's Reefer, 1979-80	15	20	____	____	____
9878	Good and Plenty Reefer, 1979	15	20	____	____	____

		Exc	LN	Color	Cond	$
9879	Hills Bros. Reefer, 1979-80	15	20			
9880	Santa Fe Reefer, orange, 1979	35	50			
9881	Rath Packing Reefer, 1979	30	40			
9882	N Y R B "Early Bird" Reefer, 1979	30	40			
9883	Nabisco Reefer, 1979	30	40			
9884	Fritos Reefer, 1981-82	10	15			
9885	Lipton Tea Reefer, 1981-82	10	15			
9886	Mounds Reefer, 1981-82	10	15			
9887	Fruit Growers Reefer, 1984	30	35			
9888	Green Bay & Western Reefer, 1983	50	70			
12700	Erie Magnetic Crane, 1987	150	200			
12701	Operating Fuel Station, 1987-88	75	90			
12702	Operating Control Tower, 1987	50	75			
12703	Icing Station, 1988	—	65			
12704	Dwarf Signal, 1988	—	18			
12705	Lumber Shed, 1988	—	6			
12706	Barrel Loader Building, 1987-88	—	7			
12707	Billboard Set, 1987-88	5	6			
12708	Street Lamps, 1988	—	8			
12709	Banjo Signal, 1987-88	—	25			
12710	Engine House, 1987-89	—	23			
12711	Water Tower, 1987	—	12			
12712	Automatic Ore Loader, 1987	—	22			
12713	Automatic Gateman, 1987-88	—	36			
12714	Automatic Crossing Gate, 1987-88	—	23			
12715	Illuminated Bumpers, 1987-88	—	6			
12716	Searchlight Tower, 1987	—	26			
12717	Non-Illuminated Bumpers, 1987	—	3			
12718	Barrel Shed, 1987-88	—	6			
12719	Refreshment Stand, 1988	—	65			
12720	Rotary Beacon (Metal), 1988-89	—	36			
12721	Illuminated Extension Bridge, 1989	—	30			
12722	Roadside Diner with smoke, 1988	—	40			
12723	Microwave Tower, 1988	—	20			
12724	Signal Bridge, 1988	—	40			
12725	Lionel Truck and Trailer, 1988-89	—	10			
12726	Grain Elevator Kit, 1988	—	22			
12727	Automatic Operating Semaphore, 1989	—	25			
12728	Illuminated Freight Station, 1989	—	30			
12729	Mail Pickup Set, 1988	—	19			
12730	Plate Girder Bridge, 1988	—	7			
12731	Station Platform, 1988	—	7			
12732	Coal Bag, 1988	—	2.50			
12733	Watchman's Shanty, 1988	—	6			
12734	Passenger / Freight Station, 1989	—	15			
12735	Diesel Horn Shed, 1988	—	30			
12736	Coaling Station, 1988	—	22			
12737	Whistling Freight Shed, 1988	—	30			
12739	Tractor And Tanker, 1989	—	15			
12740	Log Package (3 logs), 1988	—	2.50			
12741	Intermodal Crane, 1989	—	220			
12742	Gooseneck Street Lamps, 1989	—	12			
12743	O Track Clips, 1989	—	4			
12744	Rock Piers, 1989	—	7			
12745	Barrel Pack (6), 1989	—	3			

No.	Item	Exc	LN	Color	Cond	$
12746	O27 Oper./Uncoupling Track, 1989	—	13			
12748	Illuminated Station Platform, 1989	—	17			
12749	Rotary Radar Antenna, 1989	—	30			
12750	Crane Kit, 1989	—	7			
12751	Shovel Kit, 1989	—	6			
12752	Historical VHS tape, 1989	—	18			
12753	Ore Load to fit ore car, 1989	—	3			
12754	Graduated Trestle Set (22), 1989	—	13			
12755	Elevated Trestle Set (10), 1989	—	11			
16000	P R R Pass., Vista Dome, 1987-88	20	25			
16001	P R R Passenger, 1987-88	20	25			
16002	P R R Passenger, 1987-88	20	25			
16003	P R R Observation, 1987-88	20	25			
16009	Pennsylvania Combine, 1988	—	25			
16010	Dry Gulch Passenger, 1988	—	35			
16011	Dry Gulch Passenger, 1988	—	35			
16012	Dry Gulch Passenger, 1988	—	35			
16013	Amtrak Lighted Combine, 1988	—	25			
16014	Amtrak Lighted Vista Dome, 1988	—	25			
16015	Amtrak Lighted Observation, 1988	—	25			
16016	New York Central Baggage, 1989	—	25			
16017	New York Central Combine, 1989	—	25			
16018	New York Central Coach, 1989	—	25			
16019	New York Central Vista Dome, 1989	—	25			
16020	New York Central Coach, 1989	—	25			
16021	New York Central Obs., 1989	—	25			
16022	Pennsylvania Baggage, 1989	—	22			
16023	Amtrak Coach, 1989	—	23			
16023	Amtrak Lighted Coach, 1988	—	25			
16102	Southern 3-D Tank, 1987	—	35			
16103	Lehigh Valley Dome Tank, 1988	15	20			
16104	Santa Fe 2-D Tank Car, 1989	—	18			
16200	Rock Island Boxcar, 1987	—	10			
16201	Wabash Boxcar, 1988	—	10			
16204	Hawthorne Boxcar (Appliance Store)	—	75			
16208	Pennsylvania Auto Rack w/ cars, 1989	—	30			
16209	Disney Magic '88 Sears Set, 1988	—	75			
16211	Hawthorne Boxcar, 1988	—	70			
16213	Shop Rite Boxcar, 1988	—	75			
16300	Rock Island Flatcar w/fences, 1987	6	8			
16301	Barrel Ramp, 1987	—	15			
16302	Lehigh Valley Flatcar w/derrick, 1987	—	15			
16303	P R R Flatcar w/trailers, 1987	20	45			
16304	Rail Blazer Gondola, 1987	5	7			
16305	Lehigh Valley Ore Car, 1987	40	70			
16306	Santa Fe Barrel Ramp, 1988	—	15			
16307	Nickel Plate Flatcar, 1988	—	20			
6308	B N Flatcar w/ trailer (1), 1989	—	20			
6309	Wabash Gondola, 1988	5	7			
6313	P R R Gondola w/cable reels, 1989	5	7			
6314	Wabash Flatcar w/ trailers (2), 1989	—	20			
6315	Pennsylvania Flatcar w/ fences, 1989	—	20			
6317	Pennsylvania Barrel Ramp Car, 1989	—	16			
6318	L L Dep. Flatcar w/wire reels, 1989	—	20			
5400	Pennsylvania Two-bay hopper, 1989	—	25			

		Exc	LN	Color	Cond	$
16402	Southern Hopper, 1987	—	35			
16500	Rock Island, Bobber Caboose, 1987	—	25			
16501	Lehigh Valley SP Caboose, 1987	15	20			
16503	N Y Central Transfer Caboose, 1987	—	12			
16504	Southern N5C Caboose, 1987	—	35			
16505	Wabash Caboose, 1988	8	10			
16506	S F Bay Window Caboose, 1988	25	30			
16510	New Haven B/W Caboose, 1989	—	30			
16511	Penn Bobber Caboose, 1989	5	7			
16513	Union Pacific SP type Caboose, 1989	—	22			
16600	I C Coal Dump, 1988	—	20			
16601	C N Searchlight Car, 1988	—	20			
16602	Erie Coal Dump, 1987	—	18			
16603	Detroit Zoo Giraffe Car, 1987	—	25			
16604	N Y Central Log Dump, 1987	—	18			
16605	Bronx Zoo Oper. Giraffe Car, 1988	—	28			
16606	Southern Searchlight Car, 1987-88	—	20			
16607	Southern Coal Dump, 1987	—	25			
16608	Lehigh Valley Searchlight Car, 1987	30	40			
16609	Lehigh Valley Derrick, 1987	20	25			
16610	Lionel Track Maint. Car, 1987-88	—	20			
16611	San Francisco Log Dump, 1988	—	20			
16612	SOO Line Log Dump Car, 1989	—	20			
16613	Katy Coal Dump Car, 1989	—	20			
16614	Reading Cop and Hobo car, 1989	—	30			
16615	L L Extension Searchlight Car, 1989	—	25			
16617	C & N W Boxcar with end of train device, 1989	—	25			
16618	Santa Fe Track Maintenance Car, 1989	—	20			
16701	Southern Tool Car, 1987	—	45			
16707	Southern Coal Dump, 1987	—	25			
16800	Lionel Railroader Club Ore Car, 1986	50	60			
16801	Lionel Railroader Club Bunk Car, 1988	50	60			
16802	Lionel Railroader Club Tool Car, 1989	—	25			
17002	Conrail A C F Hopper, 1987	90	110			
17100	Chessie Hopper, 1988	—	50			
17101	Chessie Hopper, 1988	—	50			
17102	Chessie Hopper, 1988	—	50			
17103	Chessie Hopper, 1988	—	50			
17104	Chessie Hopper, 1988	—	50			
17107	Sclair Covered hopper, 1989	—	50			
17200	C P Rail Boxcar, 1989	—	120			
17201	Conrail Boxcar (O), 1987	—	70			
17300	C P Rail Reefer (O), 1989	—	75			
17301	Conrail Reefer (O), 1987	40	50			
17400	C P Rail Gondola w/coal load, 1989	—	40			
17401	Conrail Gondola (O), 1987	35	50			
17500	C P Rail Flatcar w/ logs, 1989	—	55			
17501	Conrail Flatcar (O), 1987	—	40			
17600	N Y Central Caboose, 1987	75	95			
17601	Southern Woodside Caboose, 1988	70	85			
17602	Conrail Caboose, 1987	70	85			
17603	Rock Island Caboose, 1988	60	75			
17604	Lackawanna Caboose, 1988	—	60			
17605	Reading Low-cup. Woodside Cab., 89	—	45			

No.	Item	Exc	LN	Color	Cond	$
17870	L C C A Boxcar, 1987	60	100			
17871	T T O S Flatcar w/Kodak & Xerox Trailer, 1985	—	85			
17872	T T O S '88 Anaconda Ore, 1988	—	50			
17873	L C C A Tank Car	—	70			
17874	L O T S '88 Milwaukee Rd. Flat., 1988	—	90			
17876	L C C A C N & L Boxcar, 1989	—	35			
17878	T T O S Magma Ore, 1989	—	50			
18000	Pennsylvania B6 0-6-0 Switcher, 1989	—	750			
18001	Rock Island 4-8-4 Steam, 1987	500	650			
18002	(785) N Y C 4-6-4 Steam, 1987	950	1100			
18003	Delaware Lackawanna, 4-8-4, 1988	—	650			
18004	Reading Pacific 4-6-2, 1989	—	350			
18090	D & R G L C C A Steam, 2-6-4, 1989	—	35			
18200	Conrail SD-40 Diesel, 1987	250	275			
18201	Chessie SD-40, 1988	250	300			
18203	C P Rail SD-40, 1989	—	295			
18205	Union Pacific GE-8 40C, 1989	—	375			
18300	Penn. GG-1 Electric, 1987	450	525			
18301	Southern F M w/Caboose, 1988	—	395			
18302	Great Northern Engine, electric, 1988	—	400			
18303	Amtrak GG-1, 1989	—	475			
18400	Santa Fe 2-4-2 Snowplow, 1987	—	120			
18401	Lionel Hand Car, 1987-88	—	75			
18402	Burro Crane, 1988	—	120			
18403	Santa Claus Handcar, 1988	—	50			
18404	San Francisco Trolley, 1988	—	100			
18405	Santa Fe Operating Burro Crane, 1989	—	100			
18406	Lionel Oper. Track Maint. Car, 1989	—	90			
18408	Santa Hand Car, 1989	—	50			
18500	Milwaukee Road, GP-9, 1987	—	225			
18501	W M NW-2 Switcher, 1989	—	275			
18600	Atlantic Coast Line 4-4-2, 1987	—	95			
18601	Great Northern, 4-4-2, 1988	—	95			
18602	P R R Steam 4-4-2, 1987	—	90			
18604	Wabash 2-4-2, 1988	—	80			
18606	New York Central 2-6-4, 1989	—	165			
18607	Union Pacific 2-6-4, 1989	—	165			
18612	Chicago & North Western 2-6-4, 1989	—	100			
18700	Rock Island Switcher 0-4-0, 1987	—	25			
18702	V & T R R 4-4-0, 1988	NRS	NRS			
18800	Lehigh Valley GP-9, 1987	—	110			
18801	Santa Fe, U36-B, 1987	—	85			
18802	Southern, GP-9 Diesel, 1987	—	125			
18803	Santa Fe RS-3 Diesel, 1988-89	—	100			
18804	SOO Line RS-3 Diesel, 1988-89	—	100			
18805	Union Pacific RS-3, 1989	—	105			
18806	New Haven SD-18, 1989	—	105			
8901	P R R Alco A pow. (price w/18902), 88	—	125			
8902	P R R Alco A dummy, sold w/18901					
8903	Amtrak Alco A powered (price w/18904), 1988	—	125			
8904	Amtrak Alco A dummy, sold w/18903					
9000	Blue Comet Dining Car, 1987	—	85			
9001	Southern Crescent Dining Car, 1987	—	85			

		Exc	LN	Color	Cond	$
19002	Penn Diner, 1988	—	50			
19003	Milwaukee Road Diner, 1988	—	50			
19100	Amtrak Baggage car, 1989	—	150			
19101	Amtrak Combine, 1989	—	150			
19102	Amtrak Coach, 1989	—	150			
19103	Amtrak Vista Dome, 1989	—	150			
19104	Amtrak Dining car, 1989	—	150			
19106	Amtrak Observation car, 1989	—	150			
19200	Tidewater Southern Boxcar, 1987	—	15			
19201	Lancaster & Chester Boxcar, 1987	—	20			
19202	Pennsylvania Boxcar, 1987	—	50			
19203	Det. & Toledo Shoreline Boxcar, 1987	—	15			
19204	Milwaukee Road Boxcar, 1987	—	40			
19205	Great Northern Boxcar, 1988	—	40			
19206	Seaboard Boxcar, 1988	—	20			
19207	C P Rail Boxcar, 1988	—	20			
19208	Southern Boxcar, 1988	—	17			
19209	Florida East Coast Boxcar, 1988	—	17			
19210	SOO Line Boxcar, 1989	—	17			
19211	Vermont Railway Boxcar, 1989	—	17			
19212	Pennsylvania Boxcar, 1989	—	17			
19213	S P & S Boxcar, 1989	—	17			
19214	Western Maryland Boxcar, 1989	—	30			
19300	Pennsylvania Ore Car, 1987	—	22			
19301	Milwaukee Road Ore Car, 1987	—	20			
19302	Milwaukee Road Hopper, 1987	—	30			
19303	Lionel Lines Hopper, 1987	50	70			
19304	Great Northern Covered Hopper, 1988	—	40			
19305	Chessie Ore Car, 1988	—	18			
19305	Lehigh Valley Ore Car, 1987	—	75			
19307	B & L E Ore Car, 1989	—	17			
19308	Great Northern Ore Car, 1989	—	17			
19309	Seaboard Covered Hopper, 1989	—	20			
19310	Lancaster & Chester Covered Hop., 89	—	20			
19400	Milwaukee Road Gondola, 1987	—	30			
19401	Great Northern Gondola, 1988	—	40			
19402	Great Northern Crane Car, 1988	—	50			
19403	W M Gondola w/coal load, 1989	—	25			
19404	Western Maryland Flat w/trailers, 1989	—	30			
19417	N Y C / T T O S T O F C Car, 1987	—	70			
19500	Milwaukee Road Reefer, 1987	—	35			
19502	C & N W Refrigerator, 1987	—	40			
19503	B & A Refrigerator, 1987	—	35			
19504	Northern Pacific Refrigerator, 1987	—	30			
19505	Great Northern Refrigerator, 1988	—	45			
19506	Thomas Newcomen Reefer, 1988	—	25			
19507	Thomas Edison Reefer, 1988	—	25			
19508	Leonardo Da Vinci Woodside Reefer, 1989	—	20			
19509	Alexander Graham Bell Woodside Reefer, 1989	—	20			
19511	Western Maryland Reefer, 1989	—	30			
19600	Milwaukee Road 1-D Tank, 1987	—	35			
19601	North American 1-D Tank, 1989	—	55			
19602	Erie Coal Dump, 1987	—	25			

		Exc	LN	Color	Cond	$
19651	A T & S F Tool Car, 1987	—	25	___	___	___
19652	Jersey Central Bunk Car, 1988	—	25	___	___	___
19653	Jersey Central Tool Car, 1988	—	25	___	___	___
19654	Amtrak Bunk Car, 1989	—	30	___	___	___
19700	C & O Caboose, 1988	—	60	___	___	___
19701	Milwaukee Road, N5C Caboose, 1987	—	45	___	___	___
19702	Pennsylvania N5C Caboose, 1987	—	50	___	___	___
19703	Great Northern Caboose, 1988	—	55	___	___	___
19704	W M Extended-vision Caboose, 1989	—	65	___	___	___
19705	C P Rail Extended-vision Caboose, 89	—	45	___	___	___
19706	U P Extended-vision Caboose, 1989	—	65	___	___	___
19709	Pennsylvania Work Caboose, 1989	—	80	___	___	___
19800	Lionelville Cattle Car, 1988	—	100	___	___	___
19801	Poultry Dispatch Boxcar, 1987	—	25	___	___	___
19802	Carnation Oper. Milk Car, 1987	—	85	___	___	___
19803	Reading Ice Car, 1987	—	37	___	___	___
19804	Wabash Hopper, 1987	—	35	___	___	___
19805	Santa Fe Boxcar, 1987	—	32	___	___	___
19806	Pennsylvania Oper. Hopper, 1988	—	35	___	___	___
19807	Penn. Smoking Caboose, 1988	—	65	___	___	___
19808	N Y C Ice Car, 1988	—	40	___	___	___
19809	Erie-Lackawanna Oper. Boxcar, 1988	—	35	___	___	___
19810	Bosco Milk Car, 1988	—	90	___	___	___
19900	Toy Fair Boxcar, 1987	—	150	___	___	___
19901	I Love Virginia Boxcar, 1987	—	35	___	___	___
19902	Toy Fair Boxcar, 1988	—	125	___	___	___
19903	Lionel Boxcar, 1987	—	30	___	___	___
19904	Christmas Car, 1988	—	35	___	___	___
19905	I Love California, 1988	—	20	___	___	___
19906	I Love Pennsylvania Boxcar, 1989	—	20	___	___	___
33000	Lionel Lines RailScope Diesel, 1988	—	275	___	___	___
33001	HO RailScope Diesel, 1988	—	200	___	___	___
33002	RailScope Diesel w/TV, 1988	—	375	___	___	___
33003	HO RailScope w/TV, 1988	—	375	___	___	___
59629	L O T S, Mil. Rd. Log Carrier, 1988	—	36	___	___	___
81487	(See 19417)					
97330	(See 9733)					
3739469	M K T 1-D Tank, TTOS, 1989	—	39	___	___	___
Black Cave	SP Caboose, 1984	5	10	___	___	___
Nabisco	Work Caboose, 1984	NRS	NRS	___	___	___
U C S	Remote Control Tracks (O), 1970	5	8	___	___	___
No Number	L A S E R Playmat	—	5	___	___	___
No Number	Cannonball Playmat	—	5	___	___	___
No Number	Gondola, 1984	6	8	___	___	___
No Number	Station Platform	—	10	___	___	___
No Number	Rocky Mountain Playmat	—	3	___	___	___
No Number	Commando Assault Playmat	—	10	___	___	___
No Number	9040-type Boxcar	—	5	___	___	___
No Number	Black Cave Playmat	—	10	___	___	___

O Gauge Classics

No.	Description	Exc	LN	Color	Cond	$
51000	Milwaukee Road Hiawatha Set (350E Locomotive, 350WX Tender, 882 Front Coach, 883 Center Coach, 884 Coach), 88	—	1200	___	___	___
51001	Lionel "#44 Freight Special", set w/51100, 51500, 51800, 51400, 51700	—	775	___	___	___
51100	Lionel Elec. Engine, originally a No. 4 (sold w/51001)					
51400	Lionel Lines Boxcar, numbered "8814" (sold w/51001)					
51500	Lionel Lines Hopper, numbered "8816" (sold w/51001)					
51700	Lionel Lines Caboose, numbered "8817" (sold w/51001)					
51800	Lionel Lines Searchlight Car, "8820" (sold w/51001)					

Standard Gauge Classics

No.	Description	Exc	LN	Color	Cond	$
323	(See 13400)					
324	(See 134011)					
3251	(See 13402)					
390E	(See 13100)					
1115	(See 13800)					
11990	L C C A D & R G Steamer, 1990	—	NRS	___	___	___
13100	"390E" 2-4-2 Locomotive, 1988	—	550	___	___	___
13400	"323" Baggage Car, 1988	—	200	___	___	___
13401	"324" Pullman, 1988	—	200	___	___	___
13402	"325" Observation, 1988	—	200	___	___	___
13800	"1115" Passenger Station, 1988	—	300	___	___	___
51900	"440N" Signal Bridge/Cont. Panel, 89	—	300	___	___	___

SECTION IV
LARGE SCALE

		Exc	LN	Color	Cond	$
38356	L O T S Dow Tank, 1987	—	NRS			
55000	RailScope Steam 0-4-0, 1989	—	325			
65024	Extra long straight track, 1988	—	3			
65049	Wide radius curved track, 1988	—	2			
65167	Remote Switch, left-hand (O42), 1988	—	26			
65168	Remote Switch, right-hand (O42), 1988	—	26			
65523	Extra long track, 1988	—	4			
82000	Straight track bundle, 1988-89	—	2			
82001	Curved track bundle, 1988-89	—	2			
82002	Straight track box, 1988-89	—	10			
82003	Curved track box, 1988-89	—	10			
82004	Wide-radius curved bundle, 1988-89	—	5			
82006	X-long straight track, 1988-89	—	7			
82007	R Switch, 1989	—	45			
82008	L Switch, 1989	—	45			
82101	Lockon, 1988-89	—	3			
82102	Rail joiner, 1988-89	—	2			
82103	Couplers, 1988-89	—	4			
82104	Water Tower, 1988-89	—	40			
82105	Engine House, 1988-89	—	125			
82106	Watchman and Shanty, 1988-89	—	50			
82107	Passenger/freight Station, 1988-89	—	90			
82108	Uncoupler, 1988-89	—	5			
82109	Brass pins, 1988-89	—	1			
82110	Lumber Shed, 1989	—	35			
82111	Freight Platform, 1989	—	75			
82112	Six-piece figure set, 1989	—	7			
82115	Vehicle assortment, 1989	—	145			
85100	Pennsylvania Steam 0-6-0	—	125			
85101	D & R G Steam 0-6-0	—	NRS			
85102	N Y C Steam 4-4-2 w/smoke, 1988	—	275			
85103	A T S F Steam 4-4-2 w/smoke, 1988	—	275			
85104	A T S F Steam 0-4-0	—	NRS			
85105	Pennsylvania Steam 0-4-0	—	NRS			
85106	Chessie Steam 4-4-2, 1989	—	275			
85107	Great Northern Steam 4-4-2, 1989	—	275			
85108	B & O Steam 0-4-0, 1989	—	120			
85109	C P Steam 0-4-0, 1989	—	140			
85114	Merry Christmas Steam 0-4-0, 1989	—	NRS			
86000	Penn Passenger Car, 1988	—	NRS			
86001	Penn Passenger Car, 1988	—	NRS			
87000	N Y C Boxcar, 1989	—	50			
87001	Pennsylvania Boxcar, 1988	—	45			
87002	Santa Fe Boxcar, 1988	—	45			
87003	Great Northern Boxcar, 1989	—	45			
87100	Union Pacific Reefer, 1988	—	52			
7101	Pennsylvania Seaboard Reefer, 1988	—	52			
7102	C & O Reefer, 1989	—	50			
7105	Seaboard Reefer, 1989	—	50			

Large Scale		Exc	LN	Color	Cond	$
87200	Operating Handcar	—	60	___	___	___
87201	Milwaukee Road Ore Car, 1989	—	35	___	___	___
87202	B & O Ore Car, 1989	—	35	___	___	___
87400	Pennsylvania Gondola	—	36	___	___	___
87402	A T S F Gondola, 1988	—	36	___	___	___
87403	New York Central Gondola, 1988	—	36	___	___	___
87405	B & O Gondola, 1989	—	35	___	___	___
87406	Southern Gondola, 1989	—	35	___	___	___
87500	D & R G Flatcar, 1988	—	32	___	___	___
87501	Pennsylvania Flatcar, 1988	—	32	___	___	___
87502	A T S F Flatcar	—	NRS	___	___	___
87503	Illinois Central Flatcar, 1989	—	30	___	___	___
87504	Union Pacific Flatcar, 1989	—	30	___	___	___
87508	Merry Christmas Flatcar, 1989	—	40	___	___	___
87600	Alaska Tank Car, 1989	—	40	___	___	___
87601	Santa Fe Tank Car, 1989	—	40	___	___	___
87700	Pennsylvania Caboose	—	40	___	___	___
87701	D & R G Caboose	—	NRS	___	___	___
87702	Santa Fe Caboose	—	40	___	___	___
87703	N Y C Caboose	—	40	___	___	___
87704	A T S F Caboose	—	NRS	___	___	___
87705	Great Northern Caboose, 1989	—	50	___	___	___
87706	Chessie Caboose, 1989	—	50	___	___	___
87707	B & O Caboose, 1989	—	40	___	___	___
87708	Central Pacific Caboose, 1989	—	40	___	___	___
87716	Merry Christmas Caboose, 1989	—	NRS	___	___	___
87800	N Y C Extended Searchlight Car, 89	—	50	___	___	___

SECTION V
COLLECTIBLES AND PAPER

		Exc	Mint	Color	Cond	$
5800	Bumper Sticker	—	1.50			
5801	Lighter	—	3.50			
5802	Lapel Pin	—	4			
5803	License Plate	—	4			
5804	Key Chain	—	4.50			
5805	Ashtray	—	6.50			
5806	Coffee Mug	—	7.50			
5807	Sport Cap	—	8			
5808	Brass Key Chain	—	8			
5809	Engineer's Glass	—	9			
5810	Sleeping Boy Poster	—	9			
5811	Nickel Plate Poster	—	9			
5812	Rail Blazer Poster	—	9			
5813	Portable Tool Kit	—	11.50			
5818	Flashlight	—	20			
5819	Pen and Pencil Set	—	20			
5820	Travel Alarm Clock	—	20			
5821	Beverage Coaster Set	—	70			
5822	Wrist Watch	—	110			
JC-1	Lionel Johnny Cash Album, 1976	3	6			
No Number	Lionel Pennant	2	4			
No Number	Lionel Clock, 1976-77	25	30			

The following is a partial list of Lionel paper from 1945-1969. For the complete listing, see *Greenberg's Guide to Lionel Trains, 1945-1969, Volume II.* Catalogue measurements are given with width followed by height.

	Good	Exc
1945		
Advance Catalogue	NRS	NRS
Consumer Catalogue 8-1/2" x 11", red & black, 4 pg. folder		
(A) Consumer original	NRS	NRS
(B) Consumer reproduction	.25	.50
Candid Camera Shots of Lionel Trains 5-7/8" x 8-3/4", 20 pgs.	2	5
Plans and Blue Prints 5-7/8" x 8-13/16", 20 pgs.	2	5
1946		
Advance Catalogue 10-1/2" x 8-5/16", red and black, 24 pgs.	30	70
Consumer Catalogue 8-3/8" x 11-1/4", full-color, 20 pgs.	5	50
Consumer Catalogue Reproduced, 20 pgs.	1	3
Consumer Catalogue Liberty, 8-3/8" x 11-1/4", 16 pgs.	5	75
Consumer Catalogue Reproduced, 16 pgs.	1	3
Instructions for Assembling and Operating Lionel Trains	2	5
Scenic Effects 6" x 9", 24 pgs.	2	5

	Good	Exc
1947		
Advance Catalogue "The Lionel Line for 1947", 14" x 11", 22 pgs.	10	75
Consumer Catalogue 11-1/4" x 7-5/8", full-color, 32 pgs.	10	35
Instructions for Assembling and Operating Lionel Trains	2	5
Fun With Lionel Model Railroading Red and black, 32 pgs.	3	5
1948		
Advance Catalogue "Lionel for 1948", 14" x 11", 20 pgs.	15	60
Consumer Catalogue 11-1/8" x 8", full-color, 36 pgs.	10	35
Instructions 5-1/2" x 8-1/2", 40 pgs. plus cover	2	5
Promotional Poster 18" x 19-1/8"	2	15
Paper City 38" x 25", six cut-and-assemble buildings	3	10
1949		
Advance Catalogue 17" x 8-1/4", 24 pgs.	35	100
Consumer Catalogue 11-1/4" x 8", full-color, 38 pgs.	35	100
Instructions 5-1/2" x 8-1/8", 56 pgs.	1	5
Lionel 3D Poster with Glasses	4	12
Track Layouts 11" x 8-1/2", 16 pgs.	1	3
1950		
Advance Catalogue Gold cover, black and white inside pages	30	90
Consumer Catalogue 11-1/4" x 8", full-color, 44 pgs.	25	75
Consumer Catalogue 11" x 8", red and black, 40 pgs.	15	55
Instructions 5-1/2" x 8-1/2", 64 pgs.	1	5
Art Prints of 19th Century Locomotives Four color prints	3	12
1951		
Advance Catalogue 11" x 8", red, 24 pgs.	15	75
Consumer Catalogue 11-1/8" x 7-3/4", full-color, 36 pgs.	10	50
Instructions 5-1/2" x 8-1/2", 64 pgs.	1	5
Romance of Model Railroading with Lionel Trains 9" x 6", 32 pgs.	3	9
1952		
Advance Catalogue 11-1/8" x 8", 40 pgs.	15	50
Consumer Catalogue 11-1/8" x 7-3/4", full-color, 36 pgs.	7	30
Lionel Accessories 9" x 7", 16 pgs.	2	10
Paper City 42" x 22", four vehicles, three buildings	5	1[illegible]
Offical Book of Rules 4-3/16" x 6-1/8", 16 pgs.	2	8
1953		
Advance Catalogue 11-1/4" x 7-5/8", 44 pgs.	15	4[illegible]
Consumer Catalogue 11-1/4" x 7-5/8", full-color, 40 pgs.	10	2[illegible]
Distributor's Advertising Promotions 8-1/2" x 11", 16 pgs.	3	1[illegible]
How to Operate Lionel Trains 8-1/2" x 5-3/8", 64 pgs.	1	[illegible]
Lionel Accessories 9" x 6", red and black, 16 pgs.	3	[illegible]
1954		
Advance Catalogue 11-1/4" x 7-5/8", 44 pgs.	10	3[illegible]
Consumer Catalogue 11-14" x 7-5/8", full-color, 44 pgs.	7	2[illegible]
Distributor's Advertising Promotions 8-3/8" x 11", 16 pgs.	2	1[illegible]
Accessories Catalogue 9" x 6", 20 pgs.	2	1[illegible]
How to Operate Lionel Trains 5-1/2" x 8-1/2", 64 pgs.	1	[illegible]

	Good	Exc
1955		
Advance Catalogue 11-1/4" x 7-3/4", black and white, 20 pgs.	10	30
Consumer Catalogue 11-1/4" x 7-5/8", full-color, 44 pgs.	6	22
Elliott Rowland Catalogue 8-3/8" x 11-1/8", pulp paper	2	10
How to Operate Lionel Trains and Accessories 8-1/2" x 5-1/2", 64 pgs.	1	5
Accessory Catalogue 11-1/4" x 7-3/4", 20 pgs.	5	20
1956		
Advance Catalogue 11" x 8", red cover, 48 pgs.	10	30
Consumer Catalogue 11-1/4" x 7-5/8", full-color, 40 pgs.	5	15
How to Operate Lionel Trains 8-1/2" x 5-1/4", 64 pgs.	1	4
Elliott Rowland Catalogue 8-1/4" x 11", 24 pgs.	2	9
1957		
Advance Catalogue 11" x 8-1/4", red and black cover, 54 pgs.	10	25
And Now HO — By Lionel 10-3/4" x 7-5/8", 4 pgs.	1	3
Consumer Catalogue 11-1/4" x 7-1/2", full-color, 52 pgs.	3	10
Accessory Catalogue 10" x 7-1/2", 32 pgs.	1	5
How to Operate Lionel Trains 8-1/2" x 5-1/2", 64 pgs.	1	4
1958		
Advance Catalogue 10-7/8" x 8-1/4", red and black cover, 64 pgs.	10	25
Consumer Catalogue 11-1/4" x 7-5/8", full-color, 56 pgs.	4	13
Accessory Catalogue 11-1/8" x 8", red and black cover, 32 pgs.	2	6
HO Catalogue 8-1/8" x 10-7/8", full-color, 6 pgs.	1	3
1959		
Advance Catalogue 8-1/2" x 10-7/8", full-color cover, 44 pgs.	7	20
Consumer Catalogue 11" x 8-1/2", full-color, 56 pgs.	4	15
Accessory Catalogue 11" x 8", red and black front cover, 36 pgs.	2	5
HO Catalogue 8-1/8" x 11", full-color, 8 pgs.	2	4
1960		
Advance Catalogue 8-1/2" x 11", color cover, 60 pgs.	5	18
Consumer Catalogue 11" x 8-3/8", full-color, 56 pgs.	5	10
Accessory Catalogue 8-5/8" x 11", 40 pgs.	2	6
How to Operate Lionel Trains 8-1/2" x 5-3/8", 64 pgs.	2	4
Lionel Track Layout for O27, O and HO Gauges 8-3/8" x 11", 4 pgs.	.50	1
HO Catalogue 8-1/2" x 10-7/8", full-color, 12 pgs.	2	4
How to Operate Lionel HO Trains 8-1/2" x 5-1/2", 24 pgs.	2	4
1961		
Advance Catalogue 8-1/2" x 11", color cover, 76 pgs.	7	20
Consumer Catalogue 8-1/2" x 11", red and black cover, 56 pgs.	1	10
1962		
Advance Catalogue 8-1/2" x 11", four-color cover, 64 pgs.	3	10
Consumer Catalogue 8-1/2" x 11", 100 pgs.	2	12
Accessory Catalogue 8-3/8" x 10-7/8", full-color cover, 64 pgs.	2	8
Accessory Catalogue 8-3/8" x 10-7/8", 40 pgs.	2	7
1963		
Advance Catalogue 8-1/2" x 11", yellow, b/w cover, 80 pgs.	3	10
Consumer Catalogue 8-3/8" x 10-7/8", pulp paper, 56 pgs.	2	6
Accessory Catalogue 8-3/8" x 10-7/8", 40 pgs.	2	5

		Good	Exc
1964			
Consumer Catalogue	8-3/8" x 10-7/8", black and blue ink, 24 pgs.	2	5
Lionel Raceways Catalogue	8-1/2" x 11", 12 pgs.	2	5
1965			
Consumer Catalogue	8-1/2" x 10-7/8", 40 pgs.	2	7
Lionel-Porter Science Sets	8-3/8" x 10- 7/8", 8 pgs.	1	3
Lionel-Spear Catalogue	8-1/2" x 11-1/8", 4 pgs.	1	3
1966			
Consumer Catalogue	10-7/8" x 8-3/8", full-color, 40 pgs.	1	4
1967			
Catalogue not issued	— 1966 catalogue was used.		
1968			
Trains and Accessories Catalogue	8-1/2" x 11", full-color, 8 pgs.	2	8
1969			
Consumer Catalogue	11" x 8-1/2", full-color, 8 pgs.	2	5

RAILROAD NAME ABBREVIATIONS

A C L — Atlantic Coast Line
A L A S K — Alaska Railroad
A T & S F (A T S F) — Atcheson, Topeka, and Santa Fe
B A R — Bangor and Aroostook
B & L E — Bessemer and Lake Erie
B & M — Boston and Maine
B & O — Baltimore and Ohio
B N — Burlington Northern
C N — Canadian National
C & N W (C N W) — Chicago and North Western
C & O — Chesapeake and Ohio
C P — Canadian Pacific
C & S — Colorado and Southern
Detroit & Mack — Detroit and Mackinac
D & H — Delaware and Hudson
D L & W — Delaware, Lackawanna, and Western
D M & I R — Duluth, Missabe, and Iron Range
D & R G — Denver and Rio Grande
D & R G W — Denver and Rio Grande Western
D T & I — Detroit, Toledo, and Ironton
D & T S — Detroit & Toledo Shore Line
E J & E — Elgin, Jolie, and Eastern
Erie-Lack — Erie-Lackawanna
F E C — Florida, East Coast
G M & O — Gulf, Mobile, and Ohio
G N — Great Northern
G T W —Grand Trunk Western
I C — Illinois Central
J C — Jersey Central
L & C — Lancaster and Chester
L C C A — Lionel Collectors Club of America
L & N — Louisville and Nashville
L L — Lionel Lines
L N A C — Louisville, New Albany, and Corydon
L O T S — Lionel Operating Train Society
L V — Lehigh Valley
M D & W — Minnesota, Dakota, and Western
M N S (M N & S) — Minneapolis, Northfield, and Southern
M & St L — Missouri and St. Louis
M P (MoPac) — Missouri Pacific
M P A — Maryland and Pennsylvania (Ma and Pa)
N C & St L — North Carolina and St. Louis
N & W — Norfolk and Western
N K P — Nickel Plate Road
N Y C — New York Central
N Y N H & H (N Y R B) — New York, New Haven, and Hartford
P C — Penn Central
P F E — Pacific Fruit Express
P H D — Port Huron and Detroit
P & L E — Pittsburgh and Lake Erie
P R R — Pennsylvania Railroad
R E A — Railway Express Agency
R F & P — Richmond, Fredericksburg, and Potomac
S C L — Seaboard Coast Line
S F — Santa Fe
S P — Southern Pacific
S P & S — Spokane, Portland, and Seattle
S U N X — Sunoco
T A G — Texas, Alabama, and Georgia
T C A — Train Collectors Association
T & P — Texas and Pacific
T P & W — Texas, Peoria, and Western
T T O S — Toy Train Operating Society
U P — Union Pacific
U S M C — United States Marine Corps
Va. Chem. — Virginia Chemical
V & T R R — Virginia and Truckee Railroad
W & A R R (W & A) — Western and Atlantic Railroad
W M — Western Maryland
W P — Western Pacific